CERTAIN ISLANDS

Some other books by David Divine

Destroyer's War

Road to Tunis

The Nine Days of Dunkirk

These Splendid Ships

The Blunted Sword

The Broken Wing

The North-West Frontier of Rome:
a military study of Hadrian's Wall

Mutiny at Invergordon

DAVID DIVINE

Certain Islands

a personal selection

Arco Publishing Company, Inc.
New York

First published in the United States in 1972 by
Arco Publishing Company, Inc.
219 Park Avenue South, New York, N.Y. 10003

Library of Congress Catalog Number 72–075364
ISBN 668–02632–4

Printed in Great Britain

Contents

List of Illustrations

This same ocean has in its western regions
certain islands known to almost everyone
by reason of the great number of those
that journey to and fro.

JORDANES: *History of the Goths*

Discovery

The special requirement of islands is to be discovered. An undiscovered island is as nebulous as an unconceived child. Essentially it is something that lies beyond the horizon – the spectral pallor of an invisible lagoon reflected in the blue of a Pacific sky, the immense baroque piling of cumulus above an unseen peak, the smoke of an unguessed volcano, a scent on a night wind.

Even the offshore islands were in the beginning intangible. To reach them, to give them reality for anything except the eye, required nothing less than the conquest of the sea. The manner of that conquest is unimportant. It could be that a hunted man found on the shore a log of drftwood, shaped like the letter Y, and, balanced precariously in its crotch, kicked his way across a strait to the refuge of the land beyond the water. It might be that a member of a river people with a developed riding log came down a tideway and was carried seaward on the ebb to an islet off the mouth. It may be that someone swam with a bundle of reeds for float. It hardly matters; what matters is that after the first crossing the sea was no longer a barrier. However slowly, with whatever set-backs, it was converted from a limit, an ending, to a beginning.

In the process of that conversion islands were a primary necessity. They became the staging posts of the conquest of the sea. At intervals they were an end in themselves, big enough for the refuge of entire tribes, rich enough for the development of peoples, capable of sustaining progress. But because they were islands reached from somewhere, the level of development of the people who inhabited them made them always capable of reaching out to somewhere else.

To compare an island with a continental mainland is not enough. It is better to contrast it. A national territory on a continent has theoretical boundaries, which can be changed, inland at least, by war or by policy. An island has finite boundaries. Its borders are sharply marked by the singular difference between earth and sea. An island is in itself a complete whole. Even if it is big enough to be sub-divided between different peoples, as Britain is, or as Haiti has become, it is still physically an identity. Because of this, island peoples differ from mainland peoples. The degree of difference varies, not necessarily in relation to the size of islands, but it exists always. Island peoples have a direct and unremitting relationship with the sea. Insularity is positive.

Because of this it was an island people of Crete who perfected the first thalassocracy – the first sovereignty of the sea – even as it was the island people of Britain who developed what will almost certainly be the last.

Because of all these factors, because of the hope of safety and the security of refuge, because of the magic of surmise and supposition, because of the undefinable mystery that lies beyond horizons, islands have been essential to the progress of modern man. Because they were involved in the very first days of that progress, they acquired early a mystical quality of their own.

Some of them still sustain it.

Greece gave us our earliest history – and our most powerful mythologies – and the history of Greece is an inextricable mingling of the discovery of islands, the influence of the mainland on islands, the employment of islands to influence the mainland. The *Odyssey*, the master book of the early expansion of the West, is threaded on an endless succession of islands which link the myths and the gods, the warriors and the kings, and all of these with the enormous unbridled fantasy of the first explorations. The insubstantiality of much of it, of its geographical system, of its distances and its directions, and most of all of its chronology, is the essential stuff of sea adventure. Overdrawn, often impossible, it is a literary compilation of spoken boastings set in immortal verse. Traditionally the seaman, harshly confined to truth in conflict with the ocean, exaggerates achievement to stun the stay-at-home. '. . . the Anthropophagi, and men whose heads do grow beneath their shoulders' were still stock-in-trade to Shakespeare.

The precise limits of the area covered by the voyages of the *Odyssey* are, largely by reason of this, still argued after three thousand years. It began at Troy, almost at the mouth of the eastern narrows of the Mediterranean. From Troy the track of Odysseus runs north-west to the head of the Aegean and then south down the length of that islanded sea. From the Peloponnese it strikes across the Mediterranean to the palm-crowded Land of the Lotus Eaters off the coast of Africa. The Laestrygonians lived traditionally at the north of Sardinia. If Gozo was the island of Calypso, as the Maltese claim, the western limit of the wanderings was roughly at the longitude of Cape Bon. The *Odyssey*, then, is a synthesis of legend and fact constructed out of the claim and counter-claim of all the early exploration of the inner Mediterranean.

None the less, it contains implicit awareness of the outer seas. A strong school believes that Calypso's island was at the Pillars of Hercules. There is no island there; but Calypso was the daughter of Atlas, who 'knows the depths of the whole sea, and keeps the tall pillars which hold heaven and earth asunder'. And he lived, according to Herodotus, at the end of the world 'where the sea is no longer navigable, in that place where are the gardens of the Hesperides, where Atlas supports the sky on a mountain conical as a cylinder'. And the Hesperides, it is agreed, lived in Oceanus in the farthest west where the sun goes down.

THE FORTUNATE ISLANDS

From thirty thousand feet above the straits that separate them from the coast of Africa the islands of the Hesperides move like hurrying sea beasts across an ocean of shadowed turquoise. It is an illusion, born of air speed; they are as anchored now as they were when Juba, King of Mauretania, named them the Canaries and added them to the Roman world half a century before the birth of Christ. Juba, according to the elder Pliny, described them as 'wrapped in fire on the ultimate edge of the earth'.

They are the perfect exemplars of the innumerable aspects of the insular principle. They were offshore islands – but they stood in ocean. They were mythological – at least to the classic peoples – but they were factual to the nomad Berbers of the coast of the Rio de Oro, who saw them in the inversion of the mirage or by the smoke and fire of the volcano of the Pico de Teide. They were occupied by a primitive people long before history. They became the insubstantial refuge of medieval saints – the isles of perpetual summer and abundance. They were the victims in a second and more violent invasion with the perverted Christianity of the Renaissance. And in the final phase of the development of the West they were the springing point of the most dramatic discovery of all – the discovery of the New World.

There are nine of them, if the islets of Graciosa and Alegranza are counted: Tenerife, the focal point of the group, with the 12,000-foot volcano of the Pico de Teide – the 'mountain conical as a cylinder'; the Gran Canaria, called so because of its heroic defence against the Spaniards; Gomera, the whistling island; La Palma; and Hierro, the western islet that was the limit of the ancient world, the speck of land through which Ptolemy drew his prime meridian. These are the high volcano islands, upthrust from the ocean floor. Lanzarote and Fuerteventura, closer inshore, rise, volcanic also, from an inshore submarine plateau. The strait that separates them from the continent of Africa is barely sixty-three miles wide.

Rome called them the Fortunate Islands – disregarding Juba's forbidding description – in what seems to have been a remembrance of Hesiod, who had written that 'Jupiter sent heroes after their death to the end of the world, to the Fortunate Islands, which are in the centre of Ocean'.

The saints rescued them. The honour belongs to Brendan, son of Finnloga. The legend that the Irish saints – and indeed the Welsh – habitually made their voyages on millstones has unhappily been dissipated by the cold approach of modern criticism. Brendan used large coracles or curraghs, and he is credited with several voyages to Brittany with which country Ireland unquestionably maintained communication. On his return from one of these voyages he was visited by a friar named Barintus – in a sense the first public relations officer of island travel – who claimed to have come from an island where the sun was invariably brilliant, the fruit perpetually ripe, and where even the birds wore crowns of gold, sang in harmony, and answered questions with an exquisite clarity. In this paradise of the sea, he said, there were no harmful insects, a matter of some importance to a medieval monk.

St Brendan built his craft of wattle and hides tanned in oak bark and buttered. He stocked it with provisions for forty days and instructed his disciples to embark in the name of the Trinity. His disciples jibbed a little, but embarked.

He appears to have headed south-east from Ireland at the start. Certainly his legend reappears in Brittany; it is even held to be original to the area by the French, who call the saint Brandaines. He continued south across the Biscay, for in due course Spain enfolds him and calls him San Borondon. From Spain he sailed south again.

The Canaries and Madeira each lay solemn claim to the honour of having been discovered by him. It is just possible of course that neither is entitled, but St Brendan sailed in a curragh, which essentially was a coastal craft, and the balance of probability rests with the Canaries. In a curragh even a saint would have had to keep close in to the coast; the Canaries are barely sixty-three miles offshore and they are high islands, visible from a very few miles out to sea. Madeira, conversely, is an oceanic island 400 miles from the African coast at the latitude of Mogador. Whichever holds the honour, St Brendan's island in the legend fulfilled all the claims of Barintus. Life on it passed in an exquisite perfection. For seven years St Brendan and his followers 'lived in the fragrance of a dream, and afterwards it seemed to have been no more than a handful of months'. At the end Brendan returned to Ireland and, despite recent experience, built a monastery at Clonfert in Galway in flat defiance of the climate. The exquisite islands of the south were re-christened The Islands of the Blest.

Brendan's story, like the *Odyssey*, is a composite of legend and exploration. Arab folk tales, Scandinavian sagas, earlier Irish stories like the tale of Maelduin, even, are drawn upon for incident. It is of no importance. Enshrined in Brendan's story once more is a record of a major voyage of exploration.

Brendan either died or went wandering perpetually in search of the floating island of Antilia, and was followed in due time by the Arabs. Abu Fatima is said to have described a voyage of the Admiral Ben Farroukh, who landed about the year 999 on Gran Canaria and found its people interested in trade. Idrisi, the Moor, geographer at the court of Roger of Sicily, a century later notes that the smoke of a magician's island was visible to the Moors of the Rio de Oro coast, and Arab maritime literature uses the name Al-Jazir-al-Khaledat, which is approximately the Fortunate Islands in Arabic.

Once more the Canaries lapsed into obscurity. The adventure of the known world was concentrated for almost 200 years upon the splendour and madness of the Crusades and there was little urge towards outward explorations. But the Crusades in their own time faltered and died with the capitulation of Acre, and chivalry in Europe was left with astonishingly little to do except to edge the Moors slowly out of Portugal and Spain.

Shortly before they were driven out of Lisbon in 1147 the Maghurin – the Lisbon Wanderers – were claimed to have rediscovered them. More certainly, in 1270 the Genoese Malocello went that way and built a castle. The Genoese flag appears on fourteenth century maps. Seventy years later a Portuguese expedition, led by another Genoese, Niccoloso da Recco, made a survey of the group, the reports of which are remarkably detailed. Just three years after, Louis de la Cerda, a

Spaniard turned Frenchman, muscled in and secured the title of King of the Fortunate Islands from Pope Clement VI. There is clear evidence that the claim was bitterly resented by the English Ambassador, who declared forthrightly that the ascription *Insulae Fortunatae* applied exclusively to Britain. The inhabitants of the Canaries were not consulted.

Long before Louis de la Cerda, however, long before the Arabs, long before Juba, long before even Homer himself, the islands were inhabited.

Thirty thousand years before Christ a great wave of invasion flowed from the east replacing the Neanderthal peoples of South-Western Europe. It brought the painters of the caves of Lascaux and the sculptors of the Aurignacian statuettes, and from the small cave near Les Eyzies in the Dordogne where their skeletal remains were first identified, they are called Cro-Magnon man.

The main wave of the flood settled across Western Europe but a part passed on slowly and remorselessly – its traces are found in Andalusia and the southern coast of Spain – elements crossed in due time from Southern Europe to Africa. Logically the crossing must have been made at the Straits of Gibraltar, for despite their astonishing development in the graphic arts, the Cro-Magnon people were Palaeolithic and an inland people at that. It is improbable that they had developed anything for the rivers and the lakes beyond perhaps a hollowed-out canoe. From Mauretania across the Straits they appear to have spread south-west into the Atlas and south down the Atlantic coast.

Certainly some of them reached the Rio de Oro. Anthropologists consider that ethnic groups still existing in the Dordogne, in Scandinavia, and in a number of areas of Spain inherit identifiable Cro-Magnon characteristics – so do the inhabitants of the Canary Islands. Perhaps as the result of a clash between these and the aboriginal African inhabitants, more probably through the sheer belly-pinch of the desert coast of the Sahara, the migration that reached the Rio de Oro looked for a refuge. There were, as has been said, islands in the frequent mirage, the smoke of the volcanoes.

A sixty-three-mile sea crossing in riverine canoes must have been an act of faith requiring either great imagination or great desperation or both. Whatever the cause, they reached Lanzerote, the first of the islands of the shallows. Hardly more than an off-lying fragment of the Sahara, Lanzerotte can have offered them small comfort – it had little water, much sand and volcanic rubble. The move south to Fuerteventura would be automatic and inevitable. This time the distance was small – barely seven miles – but Fuerteventura, the second of the islands of the shallows, must even then have been less fertile than Lanzerote. Traditionally there was only one spring on it, and crops – even the exiguous crops that were their mainstay – depended wholly on an irregular rainfall or on dew distilled in the volcanic ash.

The next step was more formidable. From the tip of Fuerteventura to Gran Canaria is almost as far as the first passage from the African coast. That had, however, been accomplished and this one followed. How long it took, how slow the making of the necessary decisions, does not matter. In course of time all the islands to La Palma and Hierro were settled. As the settlements grew, however, their isolation grew with them. There was no cohesion in the group. As they grew strong, they grew strong enough also to defend isolation.

They were, as has been said, a Palaeolithic people. Their tools were tipped with obsidian, their spears hardened with fire or pointed with horn. In Gran Canaria they used a polished battle-axe of stone. Their clubs were armed with pebbles and they carried a small shield. They had no metals, but their surviving pottery is advanced – it is classed basically as Neolithic, with their fabrics and their ornaments. Made without the wheel, it was decorated mainly with the finger-nails and it achieved a remarkable degree of excellence.

Because of their isolation the separate islands developed major differences in government; fundamentally, however, they were monarchical. Under the kings there was a system of local governorates. The titles varied from island to island, but they appear to have had substantial power. There was an elaborate priesthood, yet even the names of the gods differed between the islands. In Tenerife the Achaman had inevitably for his seat the summit of the Pico de Teide and the evil principle Guayotta lived in its deep volcanic tunnels. In Gran Canaria the supreme being was known as Atirtisma and the personal devil was Gabio. In most of the islands there were establishments, communities of primitive vestal virgins, known as the Harimaguadas. The dead of the kings and the nobility, and probably of the wealthier laity, were mummified. Limited agriculture was practised, and one of the staple articles of diet was *gofio* – grain toasted and ground with salt in a stone quern and eaten mixed with water in a dough. The flocks were mainly goats.

Stringent measures were taken in many of the islands to prevent over-population; infanticide was common to all, though the first-born appear to have been at all times spared. Intricate efforts were made to prevent irregular contacts between the sexes, and both in Tenerife and Gran Canaria pathways were doubled, women using one and men the other, to inhibit casual intercourse.

The word 'blest' has different connotations in different places.

Louis de la Cerda's consecration as King of the Fortunate Islands was singularly barren. Nothing happened until 1360, when missionaries were sent to Gran Canaria. Most of these achieved the distinguished reward of martyrdom. Not until the beginning of the fifteenth century was there an effective move. Then in 1402 Jean de Béthencourt, a gentleman of Normandy, fitted out a ship for conquest with the backing of the King of Castile.

The independence of the individual island alone made an adequate defence impracticable. Exploiting local jealousies, the invaders rapidly secured a foothold. Fuerteventura, Gomera and Hierro in turn fell, and settlement began. In course of time Béthencourt died, bequeathing his conquests to his nephew, who, despite the Castilian connection, sold them to Portugal which had made successive attempts to secure a foothold there. Other expeditions followed, the Portuguese were ousted finally by Spain, and in a long agony of defence against a sophisticated Renaissance attack a Stone Age people died. They had their triumphs but always and inevitably disaster followed upon triumph, and one by one the islands were battered to submission. It was a defence of unparalleled heroism. Against the steel of Toledo, against the crossbow and the gun of the Portuguese, the Guanche kingdoms fought for almost a hundred years until life become intolerable. In 1494, with the last coherent forces stubbornly fighting in the mountains of Tenerife, the immense and pitiful tragedy of the *modorra* began.

There has been no analysis of the *modorra* in terms of modern psychology. It was in its essence a swift, communicable and fatal melancholia. Clearly it was psychosomatic. An enormous grief seized the Guanche people and in thousands they wandered into the mountains to die. In caves and in the high places they crouched down in a paralysis for which there was no ending. Their skeletons were found in after years, huddled and brown, clustered together in an infinite pathos.

The handful that survived were baptized in the Catholic faith; resistance ended.

Three years before the onset of the catastrophe of the *modorra* three small ships lay becalmed in sight of Gran Canaria.

With their advent the islands reached the final level in the development of insularity, the moment when the people who had reached them were to reach out 'to somewhere else'. When the wind came away after three days of exasperating stillness Christopher Columbus ordered the caravel *Pinta* to detach for repairs at Las Palmas on Gran Canaria, and with the *Niña* and the *Santa Maria* stood on for Gomera. At Gomera he set in hand arrangements for completing with supplies – beef for salting, dried fish, bread, cheese, and wine. Water casks were assembled and filled. He had sailed from the Rio Tinto on 3 August, confident that he could make his last preparations in the Canaries. Tenerife, it is true, was still violently in dispute, but Gran Canaria had been in Spanish hands for almost ten years and Gomera was one of the three islands of the first conquest.

Gomera was governed by an astonishing woman, Beatriz y Bobadilla, regent on behalf of her minor son. Beatriz had been a maid of honour to Isabella of Spain; there were rumours of a liaison between her and the King, Ferdinand; Isabella, with a swift and malicious improvisation, had married her off to the brutal Hernan y Peraza, hereditary captain of Gomera; the island seemed far enough away for exile. She was absent from San Sebastian as Columbus came in with the *Niña* and the *Santa Maria* but she was expected daily. He waited on the island on the chance of buying the ship in which she was to return, but on 25 August he dropped back, impatient, to Gran Canaria to hurry on the repairs to the *Pinta*. The work had already been put in hand; on the run south the *Pinta* had damaged her rudder but by the time Columbus arrived it had already been reshipped and the damage made good. On 2 September he was back at Gomera where in the interim the *Niña* had had her lateen sails recut for conversion to square rig.

On this day he met Doña Beatriz. She was, by local legend, still young and of great beauty. It is an article of faith in the Canaries that Columbus fell instantly in love with her. His wife, the daughter of the Captain Donatory of Porto Santo in the Madeiras, had died long since; he had, however, acquired a mistress, another Beatriz, and had by her an illegitimate son. Whether at Gomera he was unfaithful to that earlier

Beatriz is not anywhere made clear. Certainly by island tradition Beatriz y Bobadilla hanged a man from a rafter of the tower of San Sebastian for suggesting that her virtue was not entirely secure. She was a woman of character.

But Columbus was a man in the grip of the *Zeitgeist*. This was the greatest era of exploration, the years in which sea voyaging moved towards a science. Columbus had prepared for his passage with the patience of a scholar. He had won his oceanic experience in the hardest school, sailing, according to a firm tradition, to Iceland and three hundred leagues beyond to the edge of the Arctic Circle. In Iceland he would have had access to the remembered sagas and to the simple outline of the Vinland legend. His marriage had given him access to charts and papers relating at least to Madeira. He had gone south with the Portuguese to the building of São Jorge da Mina, the trading fort on the Gold Coast, and he knew the latest details of the still uncompleted search for the southern route to India. It was his thesis, argued and pleaded in Portgual, begged and blustered in the court of the Catholic kings, that if a man sailed west he would find Cipango – and Cipango was Japan.

He had chosen the Canaries for his departure point not just because by his reckoning they were the closest to Cipango but because he knew the tough westerlies that blew along the possible route by way of the Azores, and he balanced this against the easy run down to the Canaries on the Portuguese trades and the steady easterlies that blew from them throughout the summer. Above all, he knew that Gomera lay almost on latitude 28 north and that all he had to do for navigation was to keep to latitude 28 and run his westing down, deviating neither to north nor to south. Cipango lay on 28 as near as evidence from Marco Polo's journals indicated.

Neither the palm trees nor the soft air of Gomera could hold him at this point of time, nor Doña Beatriz. Whatever happened in the four days that followed his return from Gran Canaria is irrelevant, whatever the island legends say is perhaps irrelevant also. Certainly the salted beef casks were stowed in the holds of the three ships, the bread was ferried aboard, the wine, the fruit and the fresh water. And on Thursday, 6 September, the squadron sailed in a light and fitful air.

There is no record of his parting with Doña Beatriz.

For three days the three ships drifted between the islands, past Hierro, always with the great peak – conical as a cylinder – of Tenerife in sight. And then on the evening of the Saturday the wind came away and at dawn on Sunday, 9 September, the ship's boy sang his morning couplet to an empty sea:

> Blessed be the light of day
> And the Holy Cross, we say.

VINLANDA INSULA

If the validity of the tradition of Gomera is accepted, there is in the sacrifice of the hurried sailing of the *Santa Maria*, the *Niña* and the *Pinta* an undertone of irony – or perhaps a gust of Homeric laughter.

America had already been discovered.

Ten years ago the claim might still have been rash. The remarkable progress of research into the early Scandinavian voyages within the decade has produced an equally remarkable advance in acceptance on the part of the scholars. The seamen had long accepted it.

The story is complex, but singularly candid. Like the Homeric poems, the sagas in the mass are an elaborate inter-mixture of religious myth and sea venturing. Unlike the Homeric poems, the two are never indissolubly entangled. This is due to three principal causes. The first, that the Scandinavian gods – neither the old gods of the *Vanir* like Njörd and Frey and Freyja, nor the later gods of the *Aesir* who controlled man's destinies with Odin and with Thor – were positively attached either to islands or even to specific places which might have been associated with human voyages; Asgard, the home of the gods, was above the world, joined to it only by the rainbow Bifröst. The second, that the great period of the writing of the sagas, when legend took permanent shape and folk memory was crystallised, was within the Christian era. The third, that the principal sea sagas are separate poems, records of a sea people whose lives were still as intimately dependent on the open ocean as with its coastal regions, and that they were formulated, sometimes with direct memory but always with a community memory of the ventures which they record, for an intensely critical and expert audience.

The Scandinavian explosion of the eighth century set up the first of the great thalassocracies of Atlantic Europe. The burning of the monastery of Lindisfarne on the coast of Northumbria in the year 793 is the effective starting point of its achievement. The withdrawal from Vinland about 1015 is its moment of declination. Between these years it compassed the greater part of the known world and more beside. From Sweden it had thrust across the Baltic and down the great Russian rivers to the Caspian, the Black Sea and Byzantium. From Denmark it cut deep into Charlemagne's Empire when Godfred, with a fleet of two hundred ships, conquered South-Eastern Friesland. In Britain and on the North Sea coasts the raid on Lindisfarne heralded the *furore Normannorum*. Considerable areas of the coast of Scotland, the islands of the Shetlands and the Orkneys, Man and ultimately Ireland itself were raided, invaded, and finally in part settled.

Norwegians, Swedes and Danes intermingled frequently in the larger forays. Much of Western France was occupied, much of the eastern coast of England overrun. The impetus of triumph carried them south down the coast of Europe. Late in the ninth century they were harassing Northern Spain. Norva passed between the Pillars of Hercules and the Norsemen called the strait thereafter Norva's Sound. In the centre of the Mediterranean their compatriots had already begun the long mercenary campaigns that ended with Roger and the Kingdom of Sicily.

The north-west voyages were a part of this same diaspora – in some degree an accidental part. The *Saga of the Faroe Islanders*, relating events that took place about the year 800, begins – sonorous even in translation:

> There was a man called Grim Kamban. He was the first man to settle in the Faroe Islands – in the days of King Harald Finehair . . .

It is perhaps not strictly true. Dicuil, the Irish geographer, said in 825 that the islands 'are now abandoned by the hermits who had sought seclusion there, owing to the arrival of northern pirates'.

The early Christian Irish had an unaccountable predeliction for the solitary life, often in the most uncomfortable circumstances. Barintus, who induced St Brendan to go south, was clearly an exception; the northern islands are scattered with monkish cells of a lamentable inconvenience, and even the Faroes were not the end.

Somewhere about the middle of the ninth century Nadd-Odd, a Norwegian outlaw who had taken land in the Faroes, was driven off course by southerly gales while returning from a visit to the homeland and came to an unknown country. He called it from its bleak white inhospitality Snowland. A little later, though the chronology is argued, he was followed by a Swede, Gardar. Also blown off course – on a voyage from Scotland – Gardar, with a positively nineteenth century approach to discovery, named the land he saw Gardarsholm. On the rumour of their discoveries Floki, another Norwegian, followed them with three ships on a planned voyage of investigation and established a settlement. Altogether, Floki wintered twice on the new island, disliked it profoundly, and returned, naming it Iceland. None the less, in the hurried exodus that developed with the harsh rule of Harald Finehair in Norway, rebels were found who were prepared to accept even Iceland. One of them, Ingolf Arnason, by tradition flung his 'high-seat pillars' – the emblems of leadership – into the sea off the Iceland coast and followed their drift to Reykjavik – the Bay of Smoke – in the year 870. With their stranding, Iceland began its progress to nationhood. The Irish hermits, it is true, were there also, but they left '. . . because they could not live alongside heathens'. The attitude has not even now been altogether abandoned.

Despite the climate, the settlement prospered.

In the year 900 a Norwegian named Gunnbjörn Ulfsson, bound towards Reykjavik, sighted the eastern coast of a new land. Like Nadd-Odd, he also had been blown off course. Unable to head up against the strong north-easterlies he overran the settlements and, far to the west, sighted a stretch of surf and rock that might have been another island. He called it Gunnbjörn's Skerry and fought his way back to Iceland.

Eighty years later Eirikr hinn raudi – better known to later years as Eric the Red – was outlawed from Iceland for certain casual slayings.

He elected to devote his three-year banishment to an exploration of the possibilities of Gunnbjörn's claim – still current in the island. In 982 he found the Skerry – and a rockbound and mountainous coast. There was no landing in the east, and Eric worked his way south to Cape Farewell and the less formidable fjords of the south-west. Here he wintered in extreme discomfort, and in the spring began a fresh exploration, naming the coast as far north as Godthaab. At the end of the third winter he returned, unabashed, to Iceland and, in what must be one of the earliest of all bogus real estate promotions, called the discovery Greenland.

It was not green except in the briefest moment of summer but on the promise of the name he gathered together an expedition, and again the settlements prospered. There was already a land hunger in Iceland, due largely to a deplorable system of land tenure. Eric found men enough willing to join him; more than 2,000 houses of the subsequent Greenland settlement have been uncovered or identified, and Eric's own farm at Brattahlid was excavated by Paul Nørlund in 1932.

The relationship between Greenland and the North American Land Mass is of importance at this point. If Greenland is an oceanic island, there is of course no argument as to priorities in the discovery of the Americas. If it is a part of the enormous shattered jigsaw-puzzle of the Canadian Archipelago, there is.

Part of the trouble in past consideration of the matter has been due to the confusion inherent in Mercator's Projection of the world map. The conventional oblong shows Greenland as a semi-continent, almost as large as South America, looming in independence at the head of the Atlantic. Geography should be taught with globes. The globe shows it as indeed a large island, three and a half times, for example, the size of Madagascar, but it shows it separated by a narrow strait less than twenty miles wide at the Robeson Channel from Ellesmere Island and Canadian territory. Godthaab, Eric's discovery towards the mouth of Davis Strait, is no more than a few score miles farther from the American mainland than was Watling Island, the landfall of Columbus! Greenland shares the American continental shelf.

It would be reasonable to argue from this that with Gunnbjörn's sighting the discovery of the New World was accomplished. It would not, however, be important. The northern venture was not yet over.

In 986 Bjarni Herjolfsson sailed for Iceland to winter with his father. Herjolf had sailed with Eric to Greenland. Bjarni proposed to follow him, though he said frankly to his crew that 'none of us has ever been in the Greenland Sea'.

He is the fourth of the accidental discoverers.

Driven south by gales and by fogs, he sighted a low-lying land covered with timber. Small hillocks fringed the shore, and he was certain that it was not Greenland.

His crew urged him to land but he refused, and set course for the north, the wind favouring them. Two days later they sighted land once more. And again the crew urged him to put in to shore for wood and water. And again he refused. There was, he said, 'no lack of either'.

After three more days they sighted land a third time, high and mountainous, with glaciers; and when for a third time the crew demanded to be put ashore, he said: '. . . that land looks ungainsome and unwinsome.'

Four days later they found Greenland, located Bjarni's father, and reported their sightings.

There is in these cool, highly professional accounts in the sagas nothing that is, in the Greek sense, 'heroic'. The Iceland writers saw their men as seamen, competent or incompetent, fortunate or unfortunate; they saw them not in relation to gods or monsters, to enchantresses or giants, but in relation to the elemental battle with the sea that they all knew.

It is this dispassion that is perhaps the primary evidence of authenticity. Ocean voyaging was never to them an adventure in itself but a means to an end. They sailed to find plunder, to conquer old land or to discover new for settlement. They did not theorize. There was no philosophical discussion as to the nature of the world or the religious significance of the sea and its winds. Bjarni's position is admirably clear. He made an uncertain start but he maintained a single objective.

The whole history of the Norwegian discoveries hangs on precisely these qualities. Always there was the determining influence of wind and sea; success was due to two things – the quality of the ships and the quality of the seamanship. Once again, some of the failure in the understanding of the Norse achievement has been due to unwarranted doubts as to these capabilities.

The ships were, in fact, entirely adequate to the voyages. The Norwegians did not, despite Victorian imagination, conquer the Atlantic with their long ships. The long ship was a war vessel – fast, light, easy to beach, capable of surprise attack and swift withdrawal, manoeuvrable in battle. The trading ships which made the Faroes voyage and its successors were bigger, sturdier, slower, but infinitely more seaworthy. The Gokstad ship, that splendid survival of the age, was built about

900. She was 76 feet long, $17\frac{1}{2}$ feet wide – a ratio of almost one to four – and by modern measurement she would be of about thirty-two registered tons. The latest finds in the Roskilde Fjord in Denmark give us magnificently the prototypes of the *hafskip*, the ocean-going ship built for the Northern trade.

There is, of course, ample evidence that there were casualties. Eric sailed on his settlement voyage with twenty-five ships – and fourteen turned back. But the proportion of loss was never high enough to discourage the venturing. They took reasonable precautions. There were even written 'sailing directions' and they were kept up to date. In 1360 Ivar Bardarson, writing admittedly later than the historic period, said:

> . . . from Snaefellsnes in Iceland where there is the shortest crossing to Greenland, the course is two days' sailing and two nights' sailing due west, and there lie the Gunnbjarnarsker, half-way between Greenland and Iceland. This was the old course, but nowadays ice has come down from the north-east out of the gulf of the sea so close to the aforesaid skerries that without risk to life no one can sail the old course and be heard of again.

The ships were capable. The voyages themselves were relatively short. Again there has been misunderstanding in the past. The discovery of North America was not a single venture but the sum of many ventures – the primary Faroes voyage, the first Iceland voyage, the Greenland voyage, and the voyage of Bjarni Herjolfsson.

Six years after Bjarni's landing, Leif Ericsson, the son of the Red, bought his ship. He tried desperately to persuade his father to lead an expedition to examine the land that Bjarni had sighted. Eric was old and tired, but after long argument he consented – and fell off his horse on his way to the ship and broke his leg.

Leif sailed with thirty-five men, probably about the year 1000.

They sighted the island of glaciers and called it Helluland – the Land of Slate. They sighted the forest coast, and Leif said: 'From its good favour this land shall have a name given to it,' and called it Markland, the land of forests. They sighted an island:

> . . . that lay north of the land and went ashore in the after boat and found dew on the grass, and put their hands in the dew and put them to their mouths and thought that they had never tasted any so sweet as it was.

Finally they sailed into the shallows of the sound that lay between the island and the ness that jutted out from the mainland, on a falling tide, and grounded. With the after boat they went in towards a river that came from a lake, and when the tide came up again they towed the ship in and made up their minds to winter there.

> There came no frost in the winter, and the herbs hardly withered, and the day was more even than in Greenland or Iceland.

In the spring Leif halved the ship's company for alternate explorations. Late in the summer Turki, the South Man, disappeared. He was found 'steep-faced and rolling-eyed, tiny-faced and miserable to look at', and when he was asked why he was late, 'he talked Dutch for a long time, rolling his eyes all ways and making faces'. When he could speak intelligibly again, he said: 'I have something new to tell. I have found vines and grapes.'

Turki deserves a place in the folklore of America, a place akin at least to that of Sethynin of the *Mabinogion* – one of the three great drunkards of the Isle of Pridain.

> In the spring when they made ready and sailed away, Leif gave a name to the land according to its produce, and called it Wineland.

Seamen, and by the middle of the twentieth century perhaps a small majority of the scholars, accepted the story of the sagas. Tangible proof was wanting.

The succession of 'finds' along the North American coast were one after another rejected. Then in 1957 an antiquarian bookseller of New Haven, Connecticut, Mr Lawrence Witten, acquired an early geographical manuscript in a nineteenth-century binding from 'a European source'.

It comprised a map of the world on thin vellum and an account of a part of the mission of John de Plano Carpini to the Mongols, long known to be missing and now named as *The Tartar Relation*. The MS. is beyond question of the middle of the fifteenth century. It is entirely accepted today as a copy by a scribe, in an Upper Rhineland bastard book hand, of an earlier manuscript. It is written on paper interleaved with vellum, which has been identified as manufactured in the Upper Rhineland, possibly at Basle, probably in about 1440.

The map is annotated, apparently by the same hand, and the reasonable assumption was that it was of the same period. Its singular importance lay in the fact that it showed not only the known medieval world but Iceland, Greenland, and a long island clearly and unequivocally labelled *Vinlanda Insula.* Its presumptive date – about 1450 – ante-dates any representation of the northern part of America on a map by almost a century and a half. Incidentally, of course, it ante-dates Columbus.

It was purchased for the Yale Library. Paul D. Mellon is reported to have assisted in the transaction, and the purchase price is reputed to have been in the neighbourhood of £100,000.

A prolonged period of scientific and scholastic examination began.

Certain problems presented themselves.

The first was the question of provenance. It became widely believed that the 'European source' was Spanish. Mr Witten, however, maintained that he was not at liberty to disclose his sources and that 'even if he could do so', according to the London *Sunday Times*, 'it would not help because the map cannot be traced back past the owner he bought it from' and once more 'I understand that the man it came from couldn't remember where he got it'. In due course the London *Observer* reported that Mr Witten had bought books from Signor Enzo Ferrajoli, a well-known Italian dealer in rare books, who was involved in the alleged illegal purchase of books from the Canons of the Cathedral of Saragossa.

The second problem was the nature of the map itself. Clearly it was a copy – but of what? No predecessor for the 'New World' sector was known. The remainder of the outline of the world, it was readily agreed, was probably derived from the map of Andrea Bianco drawn in 1436 to which apparently had been added 'the augmentation of the Indies'.

The heart of the problem was that Vikings were not cartographers. Their navigation – at least in unfamiliar waters – was traditionally by guess and by God. Two areas of the map, however, were astonishingly accurate: Iceland, which of course was completely known by the fourteenth century – though it is far more perfectly represented in this map than the mother country, Norway – and Greenland. Greenland is altogether too perfect. In the considered judgement of the late Professor E. R. G. Taylor, the Vikings never penetrated farther north than latitude 76N. But the map offers a complete outline of the north coast, and, with due acknowledgement to its simple technique, a remarkably accurate one. Even Sigurdur Stefansson, drawing his map in 1590,

140 years later, represents Gronlandia as a peninsula of an American continent.

The third problem was the worm holes. Four worm holes penetrated each of the two leaves into which the vellum of the map was folded. Four worm holes penetrated the pages of *The Tartar Relation*. But the worm holes in the two sections failed to coincide. Logically, something was missing.

In 1958, by what has been described as 'a process of mingled accident and deduction', Dr Thomas E. Marston, Curator of Medieval and Renaissance Literature at the Yale Library and *ex-officio* custodian of the Vinland Map, bought from a London bookseller a manuscript in a fifteenth century binding containing four books of the *Speculum Historiale* – Vincent of Beauvais' Mirror of History.

The MS. was written on vellum and paper, precisely in the manner of *The Tartar Relation*, the size of the pages was the same, the watermark identical, the script corresponded – and the worm holes matched.

Placed precisely in the front of a volume composed of the *Speculum* and *The Tartar Relation*, the map gave the worms a clear run through.

It is a coincidence of inestimable importance.

Map, *Speculum* and *Tartar Relation* were submitted to an Anglo-American team for investigation: Dr Marston of Yale, Mr R. A. Skelton, Superintendent of the Map Room at the British Museum, and Mr George D. Painter, the Museum's Assistant Keeper of Printed Books. In October 1965 their findings were published in a volume entitled *The Vinland Map and the Tartar Relation.*

It is a notable work of scholarship.

It examines the vellum, the paper, the scribe's hand, the map, and the worm holes. It claims that the map delineates the coast of North America as far south as the Delaware; its conclusion is summed up in one of the reviews of the book in a single sentence: 'The problem of Vinland has been solved by this one map and the Norse discovery of America established beyond dispute.'

Published with a glittering lack of tact on 11 October, twenty-four hours before Columbus Day, it lit the train to a fascinating explosion of partisan enthusiasms and scholastic incredulity. It is in part difficult to separate the two. 'This Nordic myth', thundered the chairman of the Chicago Columbus Day parade, 'makes its appearance every year.' Spain followed suit: *A.B.C.*, the monarchist newspaper, accused Yale of 'exhuming the corpse of a very old, disreputable legend . . . in an attempt to take away the merit of Spain's heroic exploit'.

The world leapt into the fray. The Poles put forward a priority for Jan of Kolno, who, they claimed, landed in Delaware sixteen years before Columbus. Warily, the Welsh suggested the rights of Madoc ap Owain Gwynneth for 1390. The Russians dealt with the map firmly. In the *Soviet Weekly* Vladimir Nevsky said: 'When you study it carefully it seems to be a clever, a rather too clever fake.' After all, the Russians had just, unaccountably, put forward the claims of the Englishman Nicholas of Lynn for 1360. The Irish remembered their monks. They had, they said, reached the Straits of Belle Isle, still seeking solitude, in the fifth century. The Chinese staked out a claim for a Buddhist monk, Profound Wisdom, a few years earlier in a spirit, presumably, of amiable rivalry. The Sons of Canaan, who had dropped from sight for a considerable while, were revived again by the Phoenician lobby for the seventh century B.C., and the Italians, countering smartly, raised once more the claims of the Etruscans a century or so earlier. The Japanese mentioned quietly the pottery of Valdivia and claimed it for 3000 B.C. and then, expressionless and oriental, overtopped everything by publishing through the Director of Penn's Museum, Philadelphia, a statement that Japanese flint blades dating back to 6000 B.C. had been found in Alaska.

Under the tumult of claim and counter-claim the scholars sharpened their pens and filled their inkwells with gall and vitriol. With that mastery of exquisite insult that is the hallmark of scholarship, somebody released the fact that the ink of the map itself differed from the ink of the manuscripts, lacking, under analysis, a content of iron. At a meeting of the Society of Antiquaries at Burlington House, London, chaired by Professor Wormald, it was declared that there were differences in the scaling and spacing of letters on the map which had led to reservations about its common origin. The legend on its left-hand corner which, the meeting was informed, claimed that 'Eric, legate of the Apostalic See and Bishop of Greenland . . . arrived in this vast and very rich land in the name of Almighty God, in the last year of our most blessed Father Pascal', was written with a different pen, so fine as to make realistic comparison difficult.

But the main weight of attack lay, as has already been described, in criticism of the delineation of Greenland. It was, said Professor Taylor, placed alongside a twentieth century map on the same scale, 'almost indistinguishable'. Scholastic politeness began to give way under the abrasive effects of controversy. Mr G. R. Crone, then Librarian and Map Curator of the Royal Geographical Society, questioning 'how such a document could have escaped notice for so long', announced firmly: 'With every respect for the erudition displayed, I consider the authors have set too high a value on this map.' It must, he claimed, have been noted in the nineteenth century if that was the date when it was removed from the *Speculum* and put in the new contemporary binding. In an admirable set-to with Mr Skelton, he finished up: 'I deplore the practice, when discussing the Vinland Map, of indulging in personal attacks on critics.' Finally, Professor Arthur Davies asserted bluntly, after outlining all the circumstances: 'These legitimate doubts point to the map as being counterfeit.'

The question of value was indeed becoming a topic of consideration. When the map was sent to Oslo for exhibition it was stated that it was insured with Lloyds for £1,500,000, a substantial sum for a sheet of vellum which, one of its sponsors admitted to the *Guardian*, could 'at an outside chance . . . be the work of a twentieth century forger', even though the other sponsor claimed: 'It would have to be the perfect forger, and as everyone knows the perfect forger does not exist.'

Rudyard Kipling once wrote a cautionary tale. Called *Dayspring Mishandled*, it is a story of a plan to discredit an authority on Chaucer. In simpler hands it might have been called 'The Layman's Revenge'. James Manallace, to break the reputation of Castorley, a leading expert on Chaucerian manuscripts, spends years studying the master's methods of establishing authenticity. He uses Castorley's own pronouncements to secure 'in minutest detail every trap he set later for his own feet'. He gets from him a recipe for the ink of the period. He uses a stone quern of appropriate date. He studies the penmanship of a particular scribe who makes particular spelling errors, and he familiarizes himself with a pen cut by that scribe in a particular manner. When he is fully expert, he forges a Chaucer manuscript – a fragment of an unknown Canterbury Tale – spoils it with deliberate misspellings, treats it, as a scribe would treat a spoil, to test new pens, folds it as a stiffener, inserts it in the back of a damaged volume, and slips the volume into a box of books destined for America.

The story has parallels that become almost alarming. The box goes to a great American library, the fragment is found and tentatively identified. Castorley is appealed to for a final decision. The MS. is sent back to England for examination, and Castorley commits himself irrevocably on the basis of his habitual tests for authenticity, all of which have been carefully provided for.

The end of it does not matter. As it stands, it is a textbook for the intricacies of literary fraud. It may owe something to the investigations of John Carter and Graham Pollard, who conducted the enquiry into the forgeries of Thomas Wise, the President of the British Bibliographical Society.

It is true that Kipling's story was published two years before the appearance of Carter and Pollard's book, but it is entirely possible that Kipling was aware of the investigations – they had been current rumour in literary circles. The story is an astonishing *tour de force* and it might, of course, by itself have inspired an imitator. Professor Davies, it is true, said flatly that 'the counterfeit was undoubtedly made when the volume was being rebound, probably after 1890'. But evidence is, at best, indeterminate.

Possibly, of course, it does not matter.

Whether the Vinland Map will eventually achieve acceptance as tangible proof of the story of the sagas has still to be determined. There is, however, other and altogether plainer evidence.

In 1960 that great Danish archaeologist Professor Johannes Brønstedt wrote in *The Vikings*:

> It would be interesting if we could trace Vinland by locating the sites of Leif's huts . . . it would be worth while conducting a systematic search for such evidence along the treeless Atlantic coast of North America by aerial survey from low-flying and slow-moving planes on the lines of those carried out by Danish archaeologists in Greenland.

In that year Dr Helge Ingstad, ex-Governor of East Greenland, and his wife, Anne Stinne, an archaeologist, decided to carry out such an investigation. Beginning at Rhode Island, they moved up the Massachusetts coast, examined Nova Scotia by aircraft and by boat, covered much of the coast of Newfoundland and Labrador, and found nothing. The following year they bought a boat of their own and, using the Stefansson map of 1590 as a basis, started to explore the estuary of the St Lawrence. Finally, after almost 4,000 miles of searching, they came to the northern tip of Newfoundland. Here, at a place called

L'Anse-aux-Meadows, they found sites which appeared to offer promise. They lay in short grassland above a shallow bay, the *anse* of the name, into which a stream of fresh water meandered. The area was exposed, preliminary trenching indicated the existence of construction work. In the four subsequent years, 1961–64, a careful excavation of eight buildings was carried out by international teams led by Anne Stinne, and including American and Canadian archaeologists, a Norwegian geologist, and other scientists. The expeditions were financed from Norway and from America.

The largest site was determined as a house about sixty by fifty feet, with a hall of considerable size equipped with a long-fire and an emberpit. It contained six rooms. At another site they found a building described as a smithy, equipped with a stone anvil and containing slag, bog-iron, and other small pieces of metal. Close to it was a pit with charcoal, possibly used in the production of iron. Amongst other finds was a spindle whorl of soapstone of a type identical with others found in Greenland.

The settlement itself strongly resembled the large number of homestead sites investigated in Greenland in recent years. Carbon 14 dating tests gave an average result in the region of the year 1000, with the latest date 1080, plus or minus – as is customary in Carbon 14 dating – seventy years.

The final assessment of the L'Anse-aux-Meadows site has still to be published. It is, however, widely accepted already as authentic, and it is possible that its foundations and the remains of its successors when they are found will provide a degree of proof for the saga narratives impossible even to the Vinland Map should its authenticity eventually be accepted.

But L'Anse-aux-Meadows and the great island of the Vinland Map share one thing at least in common – neither is the Vinland of the sagas!

The reasons for this are, in simple terms, those already offered in the examination of the voyages as proof of authenticity – the unvarnished veracity of the maritime detail; the cool, clear seaman's language in which the ventures are related.

> Two days after this they found an island that lay north of the land and went there and found dew on the grass . . .

After this landing, it will be remembered, they decided to find a place for a base camp, and they explored the sound that lay between the mainland and the island. The ship stranded there on a falling tide, perhaps because they were unfamiliar with the giant tides of the northern coast of North America, but with their small boat they explored a river that issued from a lake and, when the tide served again, they towed the ship to a mud berth. The winter was mild – 'there came no frost'. And, finally, there were the grapes.

L'Anse-aux-Meadows is not an island but only a creek on the northern tip of an island so large that it must inevitably have seemed to the first voyagers to be the mainland. The distances described are small, something that could be dealt with between tide and tide. It is, moreover, inconceivable that there was 'no frost' at the northern tip of Newfoundland even at the height of the Climatic Optimum – the warm period that had made the Greenland settlement possible. Lastly, there is no botanical evidence of grapes north of Passamaquoddy Bay on the coast of Maine at any period – and grapes are most surely essential!

These details apply in a different way to the island of the Vinland Map. If, as has latterly been claimed, it is derived not from a drawn chart of the Vikings but from an early map derived from their descriptions, it is still utterly impossible to accept it. It is, in fact, wholly at variance with the saga account. The sagas describe three places – Helluland, Markland, and Vinland. The Stefansson map shows them with a rough and ready accuracy. Nowhere in the sagas is there an iota of justification for the vast island, deeply cut into by two gigantic fjords, that appears on the map as Vinland.

The story of the sagas is entirely clear. Vinland was not the coast of North America or of Newfoundland. It was an offshore island of that coast. It was not gigantic. It was small. It was not an Isle of the Blest or yet an Island of Evil, as some islands are. It was not an Island of the Gods. It was an island whose possibilities are described on a human scale, an island of man, with his strengths and with his weaknesses.

> . . . they put their hands in the dew and put them to their mouths and thought that they had never tasted any so sweet as it was.

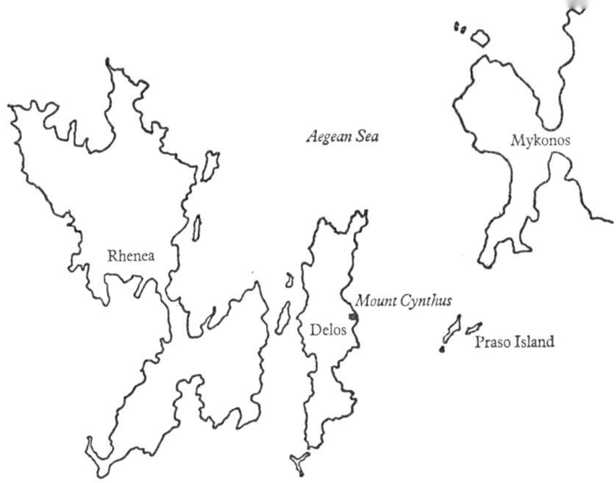

Islands of the Gods

DELOS

Delos is a golden island.

For seven months of the year the barley grass is sere, the flower stems withered and the asphodel is dry. In the bluster of the *meltemi* the surface of the island shimmers like the strings of a harp, metallic and lustrous. In the spring, for a brief time, and in the autumn the gaunt lions stand on a tapestry of flowers – purple sea stock and statice, grape hyacinths and squirting cucumber, oxalis and carmine poppies and anemones and iris. The lions have in these weeks a different splendour, and over beyond the Italian Agora the great white marble phalli acquire against the background of the flowers an almost incandescent brilliance. Then the flowers pass and Delos returns to gold.

It is pointless to compare it with other islands. To begin with, it is small even among the islets of the Cyclades – barely a full square mile. Instead of the towering magnificence of the mountain islands like Grand Canary and Hawaii and Northern Borneo its single hill, Mount Cynthus, is 368 feet high, it has no terrifying cliffs and its one palm tree is old and tired and battered. Yet it was the home of great gods, the starting point of a most ancient mythos – the birthplace of Apollo.

Before Apollo it was a nothing – a floating island – drifting unregarded in the Grecian sea. Zeus anchored it with chains of adamant to serve Leto for sanctuary when, at the end of her long flight from the wrath of Hera, she neared her time. On Delos, clinging to some infinitely remote precursor of that one palm tree or, as other scholars claim, leaning against Mount Cynthus – for she was a Titaness – Leto gave birth to Artemis and, a single day later, to Apollo.

There are variants of the story. Some of them are more understanding of Hera – she had much to put up with in the amorous idiosyncrasies of the Father of the Gods. All the variants, however, even that of the Homeric Hymn, meet at the pool of Delos.

It is never perhaps possible to subject the processes of god-making to precise analysis. By what logic, for example, can it be explained that this small island could give birth to so large a legend. Yet it happened. For most of a thousand years Mikra Dili was the navel of the religious world of the Aegean.

It may have been for more than a thousand years. With islands as with myths dates are an illusion, time is irrelevant. Certainly Delos was inhabited early in the third millennium before Christ; the hut platforms of Cynthus supply the evidence. Their buiders were a boat people, island-hopping from Asia Minor by way of Samos and Ikaria. Cretan invasions came later. Myceneans followed. Navigational problems in the Aegean had been solved early, for island succeeded island at visual distances and the open water between was navigable for craft far more primitive even than those that early Knossos used.

About 1460 B.C. the Minoan civilization collapsed, shattered by the explosion of the volcano of Santorin. The palace cities burned, the bulls and the bull dancers died. Delos passed finally into the orbit of the Peloponnese, and it may be that the tomb of the Hyperborean Maidens marks the transfer – for it is a beehive tomb, identical with those of Mycenae. The Hyperboreans were believed for long to be a people from the paradise beyond the north wind. Later interpretation claims that they were 'bringers of offerings'. If their offerings were brought to the shrine of Apollo from Mycenae, Apollo's myth must clearly date at least from the late Bronze Age. The clue, however, is tenuous.

There is no clue at all as to the length of his stay in Delos. That he was born under the palm tree and beside the pool is accepted theology. According to the Homeric Hymn, however, '. . . while still a babe, still leaping in your dear mother's arms, you slew, O Phoebus, the Serpent, the huge master of Earth'.

It was a long haul from Delos for a leaping babe. Delphi on the high shoulder of Parnassus with the cleft that was the home of the Python is 175 miles as the birds fly. The Hymn, alert apparently for possible future scepticism, provides an alternative and a more acceptable explanation. By this, Apollo in his eager youth turned himself into a dolphin and, leaping from the crest of a bright wave into a ship of Crete, began his voyage. The ship was bound for Pylos, under Nestor's palace, but the gods took charge. She made port in the gulf below Parnassus and Apollo went from there to challenge the Python. The crew of the ship went with him to become, after victory, priests of his shrine, a precedent not conspicuously followed by later generations of seamen.

Despite the move, Delos flourished. It had the absolute advantage of the birthplace. Temples rose. The shrine of Artemis, indeed, is claimed to be even older than that of her brother – it is possible that she was in fact a Mediterranean Mother Goddess – but over this the archaeologists will perhaps always quarrel. In the end three temples to Apollo stood side by side beyond the shrine of Leto. Zeus and Athena pre-empted the summit of Mount Cynthus. Isis had a temple on the terrace above the Roman city. Poseidon and the Egyptian gods established themselves, and the enormous sacred precinct was approached by a triumphal way on which the archaic and arrogant lions of Naxos stood and at the end of which there was a colossal statue of the God. The Homeric Hymn could say with certainty: 'In Delos, O Apollo, does thy heart the most rejoice.'

The association of politics and religion is many-sided and always fascinating. The importance, the wealth of Delos as a religious centre, brought it an equal importance as a political centre; by the first quarter of the fifth century, after the defeat of Xerxes, it was the headquarters of the Delian League. Its influence in that complex of jealous and irascible states was vital; its wealth was prodigious. It could and did help to build the temples of the Acropolis of Athens. Its festival – the Ionian festival – became the greatest in Greece. So immense was its sanctity that by a law of 426 death was banished from the island. The dying were taken off in boats to meet the end elsewhere, and women in childbirth, disregarding the precedent of Leto, were sent away. Even the tombs of the dead were removed, excepting only the tomb of the Hyperboreans – the graves were emptied.

The association of trade and religion is at least as fascinating. Death went but trade remained. With the collapse of Macedonia before the Roman onslaught and the destruction of Corinth, Delos reached its zenith as an entrepôt. Some of the reasons at least are obvious. From the precinct of Zeus on the top of Cynthus a circle of islands is always visible in the clear Greek light – Hecate and Little Rematia in the Dili Strait, Rheneia beyond them, Mykonos and Stapodia to the east, Tinos and Andros to the north, and then, in a vast curve so wide that the eye can scarcely accept it, Kithnos and Serifos, Sifnos and Paros and Antiparos, Naxos and Makariai and Dhenousa. It is the hub of the wheel of the Cyclades; but it is more than that: it is equidistant between Crete and Delphi, it is as far from the Peloponnese as from the Turkish shore. It is the meeting place of East and West.

To it came the ships of the nations. In its sanctity it had for long intervals peace and freedom, and trade prospers in peace and freedom. The Roman city demonstrates that with its theatre, its shops, its guest house, its rich villas, the house of Dionysius, the House of the Masks, the House of the Dolphins with their superb mosaics. There were five agoras. Its slave markets could clear 10,000 slaves in a day. It was intensely active, electric, exhilarating – and eighty-eight years before the birth of Christ, Mithradates sacked it. The priests and the merchants died, the temples were broken, the women carried off. Two hundred years later a Greek epigrammatist, speaking for the island, says: 'Ah, woe is me: how many Greek ships sail past me, Delos the desolate, whom once men worshipped.'

All this happened on an islet never more than two-thirds of a mile in breadth.

I wrote a story around these islands. It was designed for a motion picture. Its background was set in Mykonos, which today serves the empty island of the ruins. It used the white alleyways of that most lovely village of the sea, the sun on its plaster, the purple waterfall of bougainvillaea over a courtyard wall, the Square of the Three Wells, some of the 340 Byzantine chapels, most of the taverns of the waterfront and the windmills of the skyline. The story rested on Mykonos and Delos and the sea between. It was a frivolous story, light-hearted, full of laughter, basically told around a bronze statue recovered from the deep water.

Twentieth Century-Fox bought it.

They rewrote it. They injected overtones of social criticism into it. They exorcized its laughter. And after grave and careful debate they shot it on Hydra – that ravaged island over against the Gulf of Hermiones. They had, of course, to build a windmill on the ridge above the town for the sake of verisimilitude, but someone in Hollywood had been born on Hydra.

Sophia Loren looked enchanting in a wet bathing dress, Alan Ladd played the hero with, according to the *Financial Times*, 'a look of loving scrutiny that would be hard to beat in an unconscious heavyweight', Clifton Webb was impeccably Clifton Webb. One thing saved it besides Sophia. At my asking they used as model for the statue the exquisite bronze of the Boy Jockey of Artemision and they mounted him on a dolphin. I chose this in recognition of the last name of Apollo – Delphinios.

They kept the title – *Boy on a Dolphin.*

LINDISFARNE

Lindisfarne also is a golden island, but it is another gold – the pale gold of sea sand under a northern sun, wet with water of a blue that contrasts subtly with the ultramarine of the Aegean and the violent purples of the deep sea of the tropics. It also is an island of God – a different god. Because of that they call it Holy Island.

Its holiness began on the summit of a hill far inland, high above the north branch of the Tyne. There in front of Hadrian's Wall where it strides across the skyline Oswald, son of Aethelfrith, King of Northumbria, planted the cross and 'with an army, small indeed in number, but strengthened with the faith of Christ' advanced against Caedwalla.

Caedwalla died, and the old gods died with him, and Bede says: 'The place in the English tongue is called Heofenfeld – Heaven Field.'

Oswald was new to kingship. In the tough in-fighting of the northern kingdoms he had been driven for refuge to Iona, the tiny islet of the Celtic missionaries of the west of Scotland. Now, needing help to establish the faith and perhaps to prop his new-found throne, he sent an appeal to the men whom he had known. Iona sent him back a monk named Corman. Corman failed. He could do nothing with the 'uncouth Northumbrians' and Iona took thought and sent a second monk named Aidan to attempt to persuade the Northumbrians 'with the milk of mild doctrine'.

Aidan's faith in the mildness of his doctrine was no doubt complete, but with the proper caution of the Irish monks he established himself on Lindisfarne, an offshore island accessible only at low tide, and then across a waste of windy shallows.

It is not a beautiful island. It lacks the green dignity of Iona. It is sandy and flat and spray-drenched in the north-east gales. But it rises to the south-east in a rock like the prow of a ship, and twelve miles downwind from it the other islets of The Farne run like a fleet of galleys to the south.

On it Aidan built a church.

There is in the history of the northern missions a strong and sometimes faintly ribald atmosphere of rivalry. Paulinus, first Archbishop of York, assistant to St Augustine and delegate of the Church of Rome, had failed miserably in his attempt to christianize the North. Corman had failed in his turn. The incentives to success were enormous.

From the low island across the Fenham Flats spread out a masterpiece even in the astonishing story of the Christian missions. Kings and the people alike submitted. Northern England turned to Christianity, and a thousand years later another bishop, Joseph Lightfoot, could declare: 'Augustine was the apostle of Kent, but Aidan was the apostle of England.' An Archbishop of York supported the declaration. 'Bishop Lightfoot', said the Archbishop at a celebration in the ruins of the priory church on Lindisfarne, 'is not given to rhetorical exaggeration.'

It was unimportant both to Aidan and to Augustine. In their time they died, and another monk, Cuthbert, was consecrated Bishop of Lindisfarne. Two years later he also died in a tiny cell that he had built among the gulls on The Farne itself.

The importance of Cuthbert is not easy to assess. His achievement was that of a consolidator. He had been before this prior of Melrose and prior at Lindisfarne. He had left them for the tern, the cormorants and the eider duck of his beloved rock. He had been taken from The Farne at the king's behest to become bishop of the new see of Hexham and he had exchanged sees and returned as bishop to Lindisfarne. And from it he had gone back once more to his lonely cell. His temporal achievement is not apparent, his spiritual importance was enormous. Ten years after his death they opened his tomb on the south side of the altar of the church on Lindisfarne. His body showed no slightest sign of decay.

Incorruptibility is an infallible index. Cuthbert's translation to sainthood was immediate, and in its honour the priory of Lindisfarne began a book. There is a note on it in a small crabbed Anglo-Saxon hand:

> Eadfrith, Bishop of the Church of Lindisfarne, originally wrote this book in honour of God and St Cuthbert and the whole company of

the saints whose relics are on the Island. And Aethelwald, Bishop of the Lindisfarne islanders, bound it on the outside and covered it as he knew well how to do. And Billfrith, the anchorite, wrote the ornaments on the outside and adorned it with gold and with gems and gilded silver, unalloyed metal. And Aldred, unworthy and most miserable priest, glossed it in English with the help of God and St Cuthbert.

So the Lindisfarne Gospels came into being, the most important single book in the early history of the making of books in England. It is not, despite the gold and the gems and the 'unalloyed metal', glorious like the later missals and psalters and Books of Hours of France and Italy. It has no polished gold. It is not blue with the brilliance of lapis-lazuli, its colours do not burn. They are instead the colours of its own islands: the yellow of the sea grass, the reds and browns and umber of the lichen patches, the muted blues and greens of the North Sea wave, with only rarely – to point a letter or to emphasize a word – a fill of scarlet or a bar of ultramarine borrowed perhaps from sea shells and the lights of sunrise.

The four remarkable primitives of the evangelists are said to have an Italian influence. It may be so, but they could as readily be Byzantine, or they might be local portraiture arrived at by synthesis in the quiet of the scriptorium on a summer's day.

But the special quality of the Lindisfarne book is neither its portraits nor its colour but its infinite intricacy. Beyond all the customary elaboration of illumination there is in its pages a complexity of interlacing, a controlled riot of interdependence. It is at its most incredible in the five great 'carpet' pages. They are the high triumph of the ordered indiscipline of Celtic art. They are solemnly feverish; passively emotional. Academically they are described as 'anti-naturalistic'. In a sense perhaps they are, yet nature keeps breaking in. The curve of waves, the curl of a breaking sea, the sweep of wind drifting spray, and above all the sea birds. A geometrical medallion separates under scrutiny into a design of birds' heads. Their legs lengthen into elements of the interlacing, their bodies provide the central shapes, the wings elongate into endless patterns. Birds with the heads of cormorants, the wings of the black-backed gulls, the legs of waders, and the strong feet of the oyster-catchers. Sometimes there are animals – the head of a wild cat with its body prolonged to form the borders of half a page. Dogs appear at intervals, and once between two letters the small and unregarded profile of a man, cowled and dark-eyed like the saints. But always there are the birds: the birds that soared and dived and came in with backed wings to land beyond St Cuthbert's shoulders.

It is a book written in peace and fabricated in security.

A hundred years after it was written the Vikings sacked Lindisfarne. The attack of 793 was, as has already been said, the first of the wild succession of the Viking raids. The church was burned, the monastery destroyed.

The body of the saint, on the other hand, survived – and the book with it. Some of the monks survived also. Their escape is not explained. Perhaps they held out on the Beblowe Rock where, in a later age, Henry VIII was to build a fortress. Perhaps they escaped on the falling tide across the sands. It has little importance; what was important was that the inviolability of Lindisfarne was broken. St Aidan had chosen it not for defence but for peace, for isolation, and it was its very isolation that made it a target for the Norsemen. Almost certainly they attacked at high tide when they could bring their ships round into the lee of the island, close to the harbour and to the priory. Almost certainly the attack was over and the priory in flames long before men could have crossed from the mainland to their help. An offshore islet was no protection against a maritime enemy; it was a weakness. Yet the monks went back. The body of St Cuthbert was still there.

They rebuilt church and monastery, they set up life again. For just over eighty years they clung to their islet while down the coast to the Forelands and up to Scotland the long ships raided. Iona, the mother house, was burned and burned again. Ireland itself was attacked. At first the raids were isolated, forays by small groups bent on plunder; but in the later half of the ninth century they began to take a different colour – they became expeditions of colonization, they seized land, they penetrated into the interior, they established settlements. In the face of this greater threat the resolution of the monks failed.

There is no record this time of the warnings that preceded the disaster of 793; no blood dripped now from the roof of the church at York, no famine spread across the land. The monks seem simply to have come to a decision achieved in quiet consideration. They abandoned Lindisfarne, and they took with them across the golden sands the body of St Cuthbert and the great book of the Gospels.

A monastic legend says that the Gospels fell from the coffin as they struggled with it through the surf and were discovered the next day after frantic search in the shallows, unstained by salt water, and that this was, beyond doubt, a miracle.

The hegira of St Cuthbert is one of the fascinating stories of the Middle Ages. It is barely sixty miles as a gull might have flown from Lindisfarne to Cuthbert's final resting place at Durham. It took them a century and a quarter to reach it. Some of the places where they rested on the way are known; much of the journey is lost in legend, much of the legend is at least odd. They may have crossed the Roman Wall beyond Housesteads in the little gap below Cuddy's Crag – the name is held by the local people to be a diminutive of Cuthbert. They do not seem to have gone by way of Heaven Field, but certainly they must have passed through Hexham, and after seven years they reached Chester-le-Street on the road that the Romans had built to serve Hadrian's bridge across the Tyne. There they set up a new see. For a hundred years the successors to St Cuthbert were the Bishops of Chester-le-Street; then, under the pressure of the Danes once more, they moved south again, always seeking security. At Durham they found it. On the high peninsula of Dunholme in the loop of the gorge of the Wear they chose a site that was defensible. Below the hill of Dunholme – the name implies a fortification – the peninsula is barely three hundred yards across. Behind it there is a high plateau. It is the nearest thing to an island in the heart of England. There, under a chapel made of boughs, they lodged the great coffin of St Cuthbert with its incised saints and the Book of Lindisfarne beside the incorruptible body while they built the White Church, the first church of the new see of Durham, and as Durham grew, Northern England was secured in the Faith.

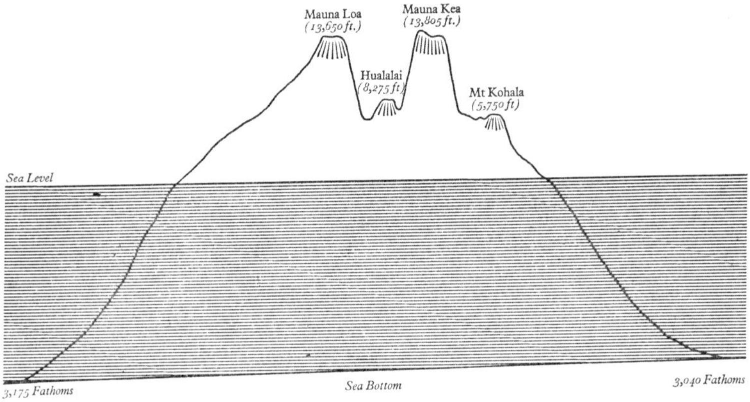

HAWAII

Hawaii – the Big Island – is three oceans distant from Lindisfarne. Its cliffs are cut of black basalt. They are enormous, lifting in great plane surfaces up to 1,500 feet above the blue of the Pacific. They follow slow curves, rising to the spaced ribs of the peak, falling to the shallow valleys that flute it with an architectural precision. Down each valley through the viridesence of the sugar cane a stream runs and at each downward curve of the cliff top leaps out towards the sea. Sometimes the water drops in shadow, sometimes the sun catches it so that against the blackness of the rock it flashes like falling crystal. From where we flew, barely 500 feet above the crests of the trade wind seas, Mauna Kea looked to us as it must have looked to Cook when *Resolution* closed the land before she bore away for Kealakekua.

It is the highest mountain in the world. Less than thirty miles offshore there is a great curve of soundings at 3,000 fathoms – sea bottom. From it the five clustered cones that make Hawaii rise in a single upward sweep. They are already 18,000 feet in height when they break surface, and above the foam that rings them they rise 13,784 feet more to the snow crest of Mauna Kea. Its crater is extinct but twenty-five miles away its twin, Mauna Loa, is at intervals vigorously active.

Cook on his third voyage, working north from his last fruitless search for that figment of the eighteenth century imagination, the Great Southern Continent, that formed the counter-balance to Asia and Europe and Africa, discovered land at Kauai. He named the long curve of islands in which it lay the Sandwich group after that compulsive card player the Earl of Sandwich, spent a bare fortnight there and sailed, loaded with coconuts and fruit, for the coast of North America. Then, carrying out a brilliant running survey, he worked his way through the Behring Straits in search of that other geographical illusion of the eighteenth century – the North-West Passage. His journal says:

> August 16. At half-past two we tacked, close to the edge of the ice . . . being then in the latitude of 70° 41′; not being able to stand on any farther.

Barred to the eastward, Cook at once fetched across to the west into the East Siberian Sea. There again the ice stopped him; it was the end of August, the season was far advanced, he stood south again through the Straits to make good his ice damage, to rest, and 'to make some improvements in geography and navigation'.

This time he made his landfall at Maui, the second of the seven large islands of the Sandwich group and, unable to find a harbour there, dropped down to the 'Big Island'. On 16 January he was off Kealakekua and sending 'Mr Bligh [of *Bounty* fame], with a boat from each ship', he began a cautious approach.

> I had nowhere, in the course of my voyages, seen so numerous a body of people assembled at one place. For besides those who had come off to us in canoes, all the shore of the bay was covered with spectators, and many hundreds were swimming round the ships like shoals of fish. We could not be more struck by the singularity of this scene; and perhaps there were few on board who now lamented our having failed in our endeavours to find a northern passage homeward.

No one in all that astonishing crowd carried weapons.

In the complex religion of the Hawaiians there were four High Gods: Ku – the War God – Kane, Kanaloa and Lono. And Lono, last of the High Gods, had one day launched his great canoes and put to sea leaving behind him word that he would return 'in a ship like a forest'. Now to the naked brown hundreds that swam about the ships he had come back. It was as simple as that.

Of the ships the tribal accounts said afterwards:

> They are islands; they are forests that have moved out into the sea.

And of their people:

> Their men are white – their skin is loose and folding – their heads are strangely shaped. They are gods, volcanoes, for fire and smoke issue from their mouths . . . Their bodies are full of treasure.

At noon the following day the decision was taken and no one knows what anxious debate had filled the previous night. From the beach the great state canoes put out.

> Their appearance was grand and magnificent. In the first canoe was Terreeoboo and his chiefs, dressed in their rich feathered cloaks and helmets, and armed with long spears and daggers: in the second canoe the venerable Kaoo, the chief of the priests, and his brethren with their idols displayed on red cloth. These idols were busts of a gigantic size, made of wickerwork, and curiously covered with small feathers of various colours, wrought in the same manner as their cloaks. Their eyes were made of large pearl oysters.

The canoes moved in procession round the ships and returned to the beach. Cook, standing in the long boat, went in astern of them.

> The king rose up, and in a very graceful manner threw over the Captain's shoulders the cloak he himself wore, put a feathered helmet on his head, and a curious fan into his hand.

Thus far it might have been no more than a greeting of great respect, the ceremonial of a state occasion, the salutation of a king. Now, however, they brought offerings and the priests wrapped him in red cloth and led him slowly to the *marae*, the sacred terrace that was the resting place of the idols. At the summit of its pyramid they presented him to the Gods. A pig that had been sacrificed in front of them was offered in turn to Cook, after which he was enthroned on a great scaffolding that towered above the rest. While a priest supported one of his arms and the king supported the other, they draped him in fresh cloth, and the priests chanted and the wind sang through the palm fronds.

The naked swimmers had vanished, the people of the villages had disappeared. This ceremonial was taboo, the rites of the return of Lono. With it this son of a farm labourer of the Yorkshire moors was elevated to the status of a deity, a naval officer became a God – a promotion altogether outside King's Regulations and Admiralty Instructions. Whether Cook recognized the depth and purpose of the ceremonial is nowhere clear. He had achieved in his earlier voyages an unparalleled record in dealing with native peoples, and it may be that he saw in the celebrations and the feasts, the dancing and the games, no more than the tribute of a friendly people. Unquestionably he used it for his needs and the needs of his ship.

Gods come expensive.

There is little doubt that in the two weeks that followed the religious ceremonies Cook unwittingly upset the economy of the island or at least that of the area about Kealakekua. It is traditional that the 'South Sea' islands flowed with milk – or at any rate coconut milk – and honey. Hawaii was no poorer than the rest; its coconuts were as rich, its taro as plentiful, its yams and sweet potatoes and bananas as good as any other. But the demands of the incessant feasting and the fresh stores for *Resolution* and *Discovery*, and above all perhaps the food for the countless visitors who came along the mountain trails and with the fleets of sightseeing canoes, exhausted the simple balance between sufficiency and dearth. Possibly they had killed off all the pigs, slaughtered the chickens and the dogs, exhausted the yam patches; certainly there were obvious signs of relief when Cook announced departure. The people of Hawaii were content now that someone else should feed their god. They saw him off with one last prodigious act of hospitality.

On 7 February, in a hard gale to the northward, *Resolution* sprung her foremast.

From this point begins tragedy, as remote, as passionless, as ordered, as it was with the Greeks. There was no harbour close at hand, and Cook perforce ran back to Kealakekua. This time the bay was empty when he put in to the anchorage. There were no swimmers. There was no singing. There were no flowers. After a long delay the chiefs came out, but the time of homage had passed. There was a sullen resentment. Cook ignored it and sent his foremast ashore to the camp that they had used before, and the carpenters started work. At once there were thefts, scuffles, temper. Ill feeling grew. On the night of the thirteenth the cutter was stolen. Early on the morning of the fourteenth Cook, following a well tested plan, went in to seize the person of a chief as hostage for its return. While the tempers rose and fell ashore, an attempt was made to get the canoes to sea past the blockade of ship's boats, and a chief of the first rank was killed. At once the women and the children were sent away, the men began to put on their war mats.

Cook had a double-barrelled musket. One barrel he kept loaded normally with birdshot; it had proved invariably effective in quelling disorder. He raised it now to deal with an aggressive individual armed with an iron spike and a stone.

The birdshot failed to penetrate the thickness of the war mat. In an instant the whole portentous edifice of deity was shattered. That the second barrel killed a man made no difference: omnipotence was

Delos, the golden island, with the stark white Lions of Naxos incandescent against sere barley grass and the faded flowers of the spring.

The 'carpet' pages of the Lindisfarne Gospels are the triumph of the ordered indiscipline of Celtic art. Into their anti-naturalistic design, however, nature intrudes with the curl of a sea, the sweep of wind and the incredible ingenuity of bird patterns.

already lost, and the mob moved in. Cook turned for an unguarded moment to call in the boats as the marines of the guard fired. With his eye no longer fixed on the people, the power of his godhead died. Stabbed in the back, he fell. He was clubbed and stabbed over and over as he lay in the warm shallows.

> Early in the morning of the 20th we saw a great number of people descending the hill, in a kind of procession . . . Eappo came into the pinnace with the bones of Captain Cook wrapped up in fine new cloth and covered with a spotted cloak of black and white feathers . . . We found in it both the hands of Captain Cook entire, which were well known from a remarkable scar on one of them that divided the thumb from the forefinger.

A powder flask had exploded in Cook's grasp in the course of his survey of Newfoundland. It seemed of no importance now.

He was not, as was believed in Britain for almost 200 years, eaten. His bones were cleaned and prepared as would have been the bones of a high chief – or a god. He was remembered. The cloaks that he had worn, the feathered idols with their eyes of large pearl oysters, were brought back to London. They lay in the British Museum for years, inadequately displayed; now they are properly housed in the new Ethnographic Gallery. There is no other record that Lono, god of the great canoes, returned – there was indeed little time left to the high gods of Hawaii.

For seven years after Cook's death, the islands were left in peace. Syphilis and gonorrhoea, which he had tried to keep from the women of Kauai in 1798, had, however, met him a year later in Hawaii, carried south by the enthusiastic eroticism of canoe travel, and the long and intricate process of depopulation was already begun. Hardly a decade after his death Kilauea, the largest active volcanic crater in the world, exploded with unprecedented violence in what must have seemed to many the final anger of the high gods. Twelve years later an ingenious Chinese introduced sugar cultivation and provoked a different explosion – less resonant, less obvious, perhaps even more destructive – of the ancient ways. In 1819 the first whaling ship dropped anchor off the coast and the whalers assumed co-operatively the blame for the effects of venereal disease. Within the next twelve months the first missionaries arrived, bringing the benefits of Christianity, clothing, and 'consumption'.

It is not easy to apportion responsibilities. By 1872 the population of the group as a whole – estimated by Cook at 400,000 – had shrunk to 56,897 of all races, native, Chinese and western. Decimated by measles and tuberculosis, the white man's passion for work, and the missionaries' passion for morality, the Hawaiian people had lost the will to live.

Lono was forgotten.

The death of Cook marks the end of the innocence of Polynesia.

Holiness is a quality not to be precisely defined. Islands have attracted it in endless succession – Malta with its malign and subterraneous goddesses in the deep cave temple of Hal Salflieni; Mont St Michel with its spired and subtle elegance; Bali with its dancing cults; Elephantine with its temples in the Nile. Islands exist in isolation. Isolation is the breeding place of mystery. Mystery is the springing point of religion. But isolation has other aspects. Mystery can be also evil. There are evil islands.

Evil

EASTER ISLAND

Te Pito o te Henua – The Navel of the World – The Eye which sees Heaven – Rapa Nui – Easter Island – it has still other names, is not in its essence evil, yet it has been overwhelmed by repetitive tragedy.

Why?

There is no reason in logic. In its geographical position it has a loneliness beyond all other inhabited islands – enough, one might reasonably assume, to ensure it from the inhumanities of man. It lies all of two thousand miles west of Caldera on the coast of Chile, 1,500 miles east of Mangareva in the Gambiers and a quarter of the width of the Pacific Ocean away from the Galapagos, the nearest islands to the northward. All its history is disaster.

The process by which man expanded across the Pacific islands has been probed more widely than any other mystery of the sea excepting only the voyage of Odysseus. It still lacks positives. Broadly it was effected by canoes – single or strutted together as catamarans. In other and more limited areas it was at least as certainly achieved by giant rafts. It advanced quite simply in accordance with canoe development, beginning with offshore island-hopping and progressing in logical order as craft and navigational skills – knowledge of the stars and the winds and the trend of the ocean swells – made possible purposeful deep-sea voyages. Its success was determined fundamentally by the character of the Pacific archipelagos, the great zones of islands that are flung, often in enormous crescents, athwart the line of the trade winds, their components widely separated but still near enough and echeloned sufficiently to reduce the awful gamble of failure to human proportions.

Its motives are less understood than any other element in the process.

Easter Island differs from all the rest. It lacks the one apparently essential attribute. It has no archipelago. The canoe peoples – for they had differing racial origins – were trained in the signs of land, trained if by nothing else by the harsh aseptic qualities of the open ocean. Even with a low archipelago they had the chance, if they missed one atoll, of sighting the blink of another in the sky. In the high islands, mountain peaks offered point after point of sighting and the clouds that lifted from them in great piles of cumulus were read like beacon smoke by the first adventurers.

Easter Island had no near atolls to reflect the shallows of lagoons against the blue of the sky; Maunga Terevaka, tallest of its score of small volcanoes, little more than 1,500 feet high, was too low to be a cloud builder.

Yet *someone* found Easter Island.

There has been as much verbal war over the manner and the credit of its finding as there has been actual war across its unhappy lava fields. Three basic theories contribute to this.

The first, whose great protagonist was the Englishman Macmillan Brown, is that its inhabitants were the last survivors of an enormous island empire. It has long been dismissed. The 1,000-fathom line lies in a circle barely ten miles from Easter Island's coast. Other islands, the ocean bottom itself, could not have subsided to that tremendous depth within the span of humanity, and the developing theory of the continental drift precludes it absolutely.

Contention today lies between the second and third theories – the Western and the Eastern. The Easterners claim that the great migrations began in South-East Asia – remotely from India, finally from the Malay peninsula. The earliest canoes worked their way along the Indonesian chain in bold primitive ventures and from its eastern end began the oceanic voyages north-east, east and south-east. Tahiti is judged to have been settled as early as A.D. 200, Samoa and Tonga by the beginning of the sixth century. Available evidence suggests that the Hawaiian chain was populated by secondary waves originating in the Society Islands and the Marquesas and spreading back south-west also to New Zealand by way of Raratonga, where they overran the earliest 'people of the land' – the *tangata whenua*. Last of all, according to the folk memories of Easter Island, sailed one small flotilla, perhaps of two canoes only, under the king Hotu Matu'a. The genealogies of the chiefs descending from Hotu Matu'a add up to suggest a twelfth century date. The great exponent of this theory in relation to Easter Island was the French ethnographist Alfred Metraux.

The Westerners put forward the contra-possibilities of migration by the great balsa-wood rafts of the coasts of Chile and Peru. Drifting with the Humboldt current almost to the equator, rafts would swing west with the equatorial current and slowly south-west again with the counter currents to the approximate latitude of the Marquesas, the Tuamotus, and possibly Pitcairn – which lies approximately on the same latitude as Easter. The ethnographic world was solidly against its practicality until Thor Heyerdahl became its champion.

Much of it still is.

The problem of Heyerdahl is that he is an empiricist, in the best sense of an abused word. He proves theories by practical experiment. In 1947, in the face of prophecies of doom, he built a balsa-wood raft following historic designs, fitted it with centre-boards – the traditional *guara* – and launched it on the Humboldt current. A hundred days later the *Kon-Tiki* stranded on the Raroia atoll.

Heyerdahl had proved a point. It was the first of a number of proofs. Deducing first from this success that the Incas also could have settled the Galapagos Islands, he made an expedition there in 1955. Thousands of shards of Inca pottery were found. A second element in the Western theory was proved. In 1955, determined to establish the rest, he led an archaeological expedition to Easter Island itself.

The fascination of Easter for scientists and layman alike lies primarily in its gigantic monolithic statues. Gods – cult figures – ancestors – there are more than a thousand of them. The smallest weighs a couple of tons. The biggest, still unfinished on the slope of the volcano of Rano Raraku, is sixty-nine feet in height.

Round the coasts of the island is a series of terraces – the *ahus* – religious platforms that have a close kinship with the Polynesian *marae*. Their earliest ritual use is uncertain but later the dead were exposed on them and in due course they were buried in them. There are more than 350, the biggest 450 feet in length. They average fifteen feet in height. Their seaward faces are sometimes of dressed masonry, sometimes of rough stone.Inland from the seaward face there is generally a paved area, occasionally as much as 250 feet across, with a platform for the dead.

Approximately 150 of these *ahus* had statues erected on them, some as many as fifteen. Their masonry divides into three classifications: the first is of large and precisely fitted stones, superbly cut; the second of carefully selected natural stones fitted with considerable pains; the third of rough dry-stone construction.

The character of the first type is the third major point in Heyerdahl's theory. Polygonal stone walling exists widely and very anciently across the world. The Temple of Apollo at Delphi stands on a terrace formed by a splendidly built wall, the stones fitted tightly to the complex angles of their neighbours. The Temple of Hagar Qim in Malta, early constructions in Crete, Italy, Spain, all contain polygonal work of varying qualities. But the style reached its highest level as an art form – clearly perfected by the love of its qualities – at Cuzco and Machu Picchu in the Andes and the great Inca cities. Additionally it was reputed to be earthquake proof and therefore of special virtue in volcanic areas.

It is Heyerdahl's contention that this Easter Island example is Inca work, that the two periods of rougher masonry that overlie parts of the great Vinapu *ahu* and other *ahus* are of later date and represent a decline from an original perfection.

Nothing is yet proven. Much still insists on explanation. The earliest Carbon 14 dating on the island recorded by Professor William Mulloy and Gonzalo Figuerao, the Chilean archaeologist, in their monumental scheme for the preservation of Easter is A.D. 857, with the usual plus or minus, for the No. 2 *ahu* at Vinapu. Skilled workers, then, were operating on the island at least as early as the ninth century and their skill bears no relation to any construction work west of them on the long approach route from the Malaysian peninsula.

The problem, as Mulloy and Figuerao bluntly admit, is controversial and complicated.

> Whether these people were Polynesians or American Indians or some sort of mixture of both, the significant fact seems to be that a probably small group, possibly reinforced by other arrivals from time to time, did appear on the island before the date indicated . . .

The statues themselves are later than the earliest *ahus*, but again dating offers immense problems. They cannot be related to anything else in the Pacific. The relatively primitive statuary of the Tuamotus provides little in the way of acceptable parallels. They are in themselves fantastic. No art form of comparable perfection has been developed elsewhere in the world from such miniscule resources or in so minute a space – Easter Island is less than fifteen miles from west to east. Their size – the great one on Rano Raraku would have been, as Heyerdahl points out, 'as high as a seven storey house' – was a crucial part of the basis of Macmillan Brown's belief. Who could have carved them? What vast labour was necessary? How were they moved? How erected?

Something of these questions also was solved empirically. Heyerdahl simply asked Pedro Atan, the 'mayor' of the little town of Hangaroa on the island, if he could carve a statue and Atan, prepared always to answer any question favourably, said 'Yes'.

With five other men, singing what he claimed were the traditional working songs, observing what he claimed were the prescribed rituals, Atan began to cut into a rock face of the quarry site. In three days the photographs show a recognizable outline cut in depth in the volcanic tuff, using only the stone picks left lying in thousands by the original sculptors in the quarry area. With different methods and different estimates of work accomplished, Atan and Heyerdahl came to the same conclusion – that a statue of medium size could be cut in twelve months by six men!

There remained two other problems – transport and erection.

The majority of the statues were carved on Rano Raraku. Some of them were transported as much as ten miles. The island had no draught animals, no machinery, no adventitious aids except only the simple lever. The precise weight of the giant statue on Rano Raraku has still to be properly calculated, but at least one fifty-ton statue was moved to its coastal *ahu*. How?

Heyerdahl solved this problem empirically also. He staged a feast. At the end of it, with the assistance of the invaluable Atan and 180 guests, a twelve-ton statue was hauled across the plain. When the ropes broke, there was laughter.

There have been innumerable explanations of the methods originally used. They range from the ability of the statues themselves to 'walk' to terrestrial magnetism. A system of giant sheer legs rocking them forward 'inches at a time' on cradles made of forked timbers has favour at present in Chile. But extrapolating from Heyerdahl's twelve-ton statue and his 180 guests, it seems entirely possible that by using timber cradles, rounded pebbles in bad patches, and perhaps seaweed as a lubricant, they could have been moved without supernatural intervention.

Atan quite certainly demonstrated that it was possible to erect them with relative ease. Twelve men, working with rough levers, raised a fallen statue on the wall of the *ahu* at Anakena. The trick has the simplicity of genius. With the levers one end of the statue was lifted a few inches and small stones were thrust into the gap. The levers were moved to the other end and raised that a fraction. More small stones were introduced. Slowly, patiently, and with the 'foreman' always keeping the stones packed beyond the danger of a slip, the statue rose in the air. When it was high enough, it slid on to its pedestal. It is possible that the great red topknots that crowned the statues offered a greater problem. Almost certainly it was solved by similar means.

Scientific method will solve, in due course, most of the outstanding problems – it may not solve everything. No one has yet approached the heart of the greatest mystery of all – the mystery of why this astoundingly vigorous entity collapsed. The broken memories of a broken race that survive on Easter Island today only tell of war in the abstract. Mayor Atan told Heyerdahl elaborate and convincing detail of the Götterdämmerung of an aristocracy – the Long Ears – of whom he was the senior surviving representative. Overwhelmed by the Short Ears in the blazing ditch at the foot of the Poike volcano, the rule of the Long Ears ended, and a civilization crashed. Heyerdahl dovetailed the story with careful economy into his basic theory. The Long Ears were the descendants of the Inca component of a dual settlement of the island; the Short Ears of the Polynesian. The Long Ears were the driving force in its stupendous development – the master race, the priests, the artists. The Short Ears supplied the labour force. In the inevitable outcome of separated communities, the Short Ears rebelled.

This outline was summarily rejected. The names of the two participants in the tragedy were, it is asserted, derived from mistranslations. There was no evidence of an aristocratic overlordship. The ditch of the Long Ears was an essay in banana plantation. The evidence of fire in the ditch was evidence of the burning of banana trash. The entire story was a splendid efflorescence of latter-day imagination designed to establish for Mayor Atan a place in history.

Possibly the sceptics are correct. Yet the alternative theories put forward scarcely supply conviction. According to these the island became overpopulated. The initial quarrels began in land disputes and brawls over fishing rights and developed into a struggle for subsistence which ended in cannibalism.

The difficulty is that one explanation is as simplistic as the other.

Both ignore certain basic considerations. At the moment of the outbreak, whatever its cause, the evidence is that Easter Island had reached its pinnacle of prosperity – in island terms. The *ahus* at the edge of the ocean provided the focal point of each separate community. Inshore the great paved areas in the shadow of the statues dominated the communal life. Beyond them the boat-shaped stone foundations of the priests' houses completed the sacred precinct. Still further inland the individual houses stretched, diminishing in density, but each with its simpler stone foundations, its stone chicken houses, its circular walled enclosures for banana cultivation and for paper-mulberry trees. Each village held its individual slice of the island cake.

Mulloy and Figuerao's survey presents evidence of a prosperous whole, the separate communities preserving a high degree of independence. There is no evidence of overcrowding. Population numbers are another of the vexed problems of Easter Island. Mulloy cautiously suggests 'a few thousand'. Present Chilean estimates go as high as 17,000; the figure seems implausible. Whatever the number, there is no evidence of the infinitely slow process by which a whole people arrives at a final crisis of food supply. The number of statues in process of being hewn from the volcano at the moment of disruption, the number on the 'finishing off' slopes of Rano Raraku, the number abandoned on the complex of roads – parts of them paved – from the quarry to the *ahus* indicate rather an immense and vibrant energy, an ample labour force, and a high degree of overall control.

Is another alternative possible?

Acknowledging that something at least of the final crescendo of statue carving should be attributed to intercommunity rivalries – or to simple ostentation – acknowledging equally that land quarrels, fishing rivalries, the absurd symptoms of miniscule 'nationalisms' combined to play their parts, there is in this suddenness of the collapse evidence of something altogether different. It has the appearance of a collapse of authority, and in the particular circumstances of Easter Island it is difficult to conceive of an overall authority that was not, in effect, religious. Was the 'war of the Long Ears' in fact a war of religion? Did the mass of the people suddenly lose faith in the giant statues and the *mana* that emanated from them over their own community segments?

Certainly in the war that followed, the overthrowing of a rival community's statues was the ultimate expression of victory. With the war, evil entered into Easter Island.

On Rana Raraku the work ceased. In the quarry where the red stone topknots were cut and hauled incredibly over the lip of the crater, work ceased. On the roads where the statues were being hauled, work ceased. And from the village communities the people drifted by degrees underground into the infinitely complex system of volcanic tunnels, caves and blow holes that seams the island. On the surface began a process of raid and brutal massacre that culminated in an appalling development of cannibalism – not the cannibalism of prowess, the symbolic reward of distant raiding against potential enemies, but the cannibalism of appetite between indigenous and related members of a single community.

On Easter Sunday in 1722 Jacob Roggeveen, the Dutch explorer, discovered the island. The discovery, contrary to experience elsewhere, brought no evil with it. Evil was already there. Statues were still standing on the *ahus* but many had already been overthrown. Roggeveen's estimate of numbers on the island was low, but numbers are perhaps irrelevant. Then and during subsequent European visits it is certain that most of the women and children stayed below ground.

Cook arrived at Easter Island in 1773. The number of statues standing was smaller still. He has relatively little to say of the islanders, but William Hodges' painting of the island with its two gaunt statues in the foreground and a dark mass of storm cloud over the sea is perhaps the most evocative memorial to its all but vanished race that remains.

The explorers were humane. White civilization was not. On an island grievously maimed by inherent savagery, the American sealer *Nancy* inflicted the first external wounds. In 1808, after a short but bloody fight, the *Nancy* carried off twelve men and ten women. Three days' sail from Easter they were released from their chains. At once they jumped over the side, swam clear of the ship, held a brief colloquy, decided on the direction of Easter Island, and began to swim towards it. Boats were lowered to pick them up but they dived and evaded them and swam on towards the horizon. There were other incidents, other raids.

Ultimate disaster came with the discovery of the rich deposits of guano on the island off Peru in 1859. In 1862 a small fleet of ships arrived off Easter. The people on the landing beaches were invited on board. They were seized at once and chained. When no more came, seamen went ashore and drove those who remained above ground with gunshot to the beaches. A thousand natives – probably half the remaining population – were taken as slaves and sold to the guano islands.

Civilization had discovered Easter Island on Easter Sunday; the great raid sailed on Christmas Day. After a long interval news of the atrocity reached the outside world, vigorous protests were made, and finally the Peruvian Government ordered their repatriation. Barely a hundred still survived. Eighty-five of these died on the passage home; fifteen returned to Easter Island, carrying smallpox with them.

It spread instantly, even into the caves of safety. At the farthest end of the island, in the all but inaccessible grotto of Ana o Keke, the children who were to be the *neru* virgins at the annual festival were immured, bleaching their skins. The smallpox spared them, but the priests who should have fed them died of it and in the terrible darkness the children starved. When the disease passed there were barely a hundred souls left on the island.

Missionaries heard of the horror and came to Easter. They suffered, left and returned. Sheep farming was introduced. Then, in a great conflict of interest between the natives and the newcomers, the Church authorities determined to move the entire population to the Gambiers. Some, despite the Church, were forcibly detained on Easter; some were enticed to Tahiti; some reached the Gambiers. Today the population is mixed, negative, and leaderless. It numbers in the region of 1,500, descended from the handful that survived the smallpox and the later handful that survived the evacuations. It has forgotten its primal innocence.

Yet, curiously, in a distracted world it has hope. The survey made by Mulloy and Figuerao for UNESCO demonstrates with a rare combination of practicality and archaeological acumen that alone among the individual 'civilizations' of the past that of Easter Island can be restored almost in its entirety.

Archaeological Easter Island has survived for two reasons: the fact that its population for practical purposes eliminated itself; and the fact that, because of its remoteness, it has not been overlaid by subsequent 'civilizations'. It is, moreover, a 'stone' development, not a timber or a metal one. Restoration primarily means the repair of damage created by internecine war – and left to the wind and the dust ever since. The authors of the plan visualize it in simple terms, using local labour and the absolute minimum of machinery – one crane on a light vehicle, one crane capable of a lift of about fifty tons.

With these, large blocks of stone, the statues themselves and, above all, the red topknots can be replaced. Heyerdahl proved a method for the smaller statues. Time and labour demand mechanics for the larger ones. Mulloy and Figuerao have shown the possibilities with the Akivi *ahu*. The restoration of the statues there is a clear indication of what can be done by remarkably simple and economic means.

If the support essential to the venture is forthcoming, Easter Island's remote and specialized contribution to the development of man is assured.

HAITI

The Windward Passage leads to the north-east like a gateway opening between the islands from the steam of the Caribbean to the cool sanity of the Atlantic. It is fifty miles wide, but in the clarity of the West Indian sun its structure is apparent through a full day's sail even with a favourable wind. Cuba, to the westward, is carved most delicately of jade. Hispaniola, to the east, is built of sculptured turquoise that changes to lapis-lazuli as the changing sun deepens the shadows of the Haitian ranges. It is serene and still and exquisite. On its shore has been concentrated more genocide, more massacre, murder, tyranny and revenge, more burning, rape and human misery than upon any comparable area of earth. It began with Columbus and the wreck of the *Santa Maria*. It has not yet ended.

Columbus made his landfall in the New World well to the northward at the island he named San Salvador in the Bahamas. It is agreed by the experts that San Salvador was what is now known as Watling Island. From it he stood south through the Bahamas group, moving from cay to cay in a triple search for 'Cipangu', for China and for the source of the gold that the gentle Arawaks of Watling Island had worn in their nostrils. From the disjointed chain of the Ragged Island Range he ran down-wind to the high land of Cuba, certain that he had found Japan.

Cuba yielded neither the Japanese Court, nor the Great Khan, nor the source of gold, and he moved east again, exploring the coast until he reached across the Windward Passage and anchored in a deep harbour under Môle St Nicolas on the island that the Arawaks called

Haiti – The Mountainous. Beating his way from there through the Tortuga Channel against the trades, he decided, implausibly, that this high green country resembled Spain and he called it *la isla Española* – the Spanish Island. On Christmas Eve, still in the pursuit of gold, he left the beautiful anchorage at Acul for the village of the cacique of Marien. He had been told that the gold mines were in the central area of the island.

Of his three ships, the *Pinta* had left him – Martín Pinzón had slipped off with a singular insubordination in search of an island called Babeque where gold could be gathered by candlelight. The *Niña*, however, was still in company; with her the *Santa Maria* was becalmed off the Limonade Pass.

The Admiral turned in, according to the journal, at 11 p.m. For forty-eight hours he had gone without sleep. No reason is given for this, but the *Santa Maria* had lain at anchor the previous night in the bay at Acul. Morrison, the American naval historian, suggests that 'celebrations' with the natives had gone on all night; others before have guessed at the possibility. Columbus himself wrote:

> It pleased Our Lord that at midnight, while I lay in bed, with the ship in a dead calm and the sea as peaceful as the water in a cup, all went to sleep leaving the tiller in charge of a boy. So it happened that the swells drove the ship very slowly on to one of those reefs . . .

It is to be doubted that Our Lord was pleased.

They manned a boat to tow her off, and the boat's crew panicked. They laid out kedges, but the ground swell worked her on to sharp heads of coral, and the tide failed to lift her. In a little it was altogether hopeless. The *Santa Maria* was abandoned.

They landed her equipment and the stores that she carried, they put her guns ashore, and from her timbers they set up the first western building in the New World – a fort founded on an error of nautical judgement, the Lord and the ship's boy assisting.

They did not yet acknowledge that it *was* a new world. Columbus still believed that if it was not Japan it was at least the spreading archipelago that lies to the southward down the China Sea. But he had now no options. The *Niña* had sailed from Spain with twenty-two men. It was not possible for her to carry the whole of the crew of the *Santa Maria* in addition to her ordinary complement. Thirty-nine men were left behind to establish a settlement in the fort that Columbus had called devoutly Villa de la Navidad, for it had been conceived on Christmas Day.

He had no anxiety for its garrison; the Taino Indians, he had written:

> love their neighbours as themselves; their conversation is the sweetest imaginable; their faces always smiling; and so gentle and so affectionate are they, that I swear there is not a better people in the world.

With an astounding energy, he was back before the end of November at the head of a relief expedition. Even so, he was too late. Navidad was burned, its people were dead. Under the leadership of Gutiérrez, the Crown's representative at the fort, the garrison had formed a plundering band to range the island for gold and women. At Maguana, in its centre, 'the sweetest imaginable' people had at last stopped smiling. Despite his guns, Gutiérrez and his men were justly and reasonably expunged. Navidad, left with a garrison of eleven men, was overrun and the last of its survivors hunted down. Columbus reached its ruin charged to christianize the inhabitants of the islands, 'to honour them much and to treat them well and lovingly'.

The contra process was already too far advanced – Columbus had reckoned without his Spaniards. The Tainos, the Haitian branch of the great Arawak nation, were indolent and gentle, a people who spent the noonday in hammocks, who fished with the suckers of captive *remora*, and whose dogs had no bark. They had just begun to learn, under the spur of Carib raiding, to defend themselves at the island level of war. They could, after the one heroic gesture against Gutiérrez, do nothing against the cold brutalities of Spain; they could not protect themselves against the catalytic effect of the knowledge of the existence of gold.

Estimates as to their numbers vary. Figures of two to three million for the four main islands of the Greater Antilles have been given, but Columbus's own estimate for Hispaniola was 250,000. Possibly it is unimportant. In 1508 a census indicated that there were barely 60,000 still left alive. By 1558 there were fewer than 500.

Thus within the brief span of eighteen months the first strands of the pattern of the 'other world' were woven: the superimposition of a sophisticated people upon a primitive race; the consequential extinction of an existing social structure. Columbus had come armed with what Leif Ericsson had lacked – gunpowder. There remained the third strand of the pattern: slavery.

Columbus has been blamed for this too. Certainly he played a part in it. In March 1494 he sent Anton de Torres, Captain of the *Mariagalante*, to carry a progress report to their Catholic Majesties.

> Item. You will tell their Highnesses, that for the good of the souls of the said cannibals, and even of the inhabitants of the said island, the thought has suggested itself to us that the greater the number that are sent over to Spain, the better.

True and pious, Columbus had a proper horror of cannibalism, but later the report changed tone a little; explaining the need for cattle and other goods, he said:

> . . . and the latter might be paid for with slaves taken from among the Caribbees, who are a wild people fit for any work.

What he was aiming at plainly was a slave export. He had become acquainted with the trade at São Jorge de Mina. Explicitly, he proposed to carry out blackbirding on a substantial scale with open boats.

The scheme was rejected. Where the early chapters of the report had been received by their Majesties with a staccato 'He has done well', the comment on this was 'Consideration suspended'.

It never reached fruition. The Caribs that he had sent over with the report were baptized and sent back; whether their cannibalism had succumbed to a Spanish diet is not clear. Long before a trade could be developed the gold mines of Cibão were suffering from a shortage of labour. To meet it a slave trade began in the opposite direction: first with the transfer of African slaves from Portugal and Spain, then by direct traffic with the Gulf of Guinea. There were unlimited Africans to make good the destruction of the gentle Taino. So was woven into the future of the New World the third strand of its pattern.

Firearms, hunting dogs, brutal treatment in the gold mines, starvation and despair exterminated a nation in simple genocide, yet by the early eighteenth century if not before, Haiti was the most prosperous colony in the New World.

It was, however, a French colony. The Spaniards had abandoned it for the mainland, for the Conquests, for the continuing search for gold. Frenchmen from the riotous pirate island of Tortuga had taken over and achieved recognition at the western end. In the year of the Bastille the population of Haiti was 550,000; there were 32,000 French, 24,000 mulattoes of varying grades, and half a million slaves.

In 1791 the slaves revolted. Under a Jamaican negro named

The forbidding cloud that overshadows Easter Island in William Hodges's dramatic landscape was prophetic. Within a hundred years Easter Island was overwhelmed in tragedy.

Yearly Venice was married to the sea. Canaletto's brilliant composition of the readying of the *Bucintoro* is one of the most sumptuous pictures of the Venice that was – and is.

Boukman, a giant and a Houngan of the Voodoo cult, there was an astonishing gathering in the forest, sacrifices were made, oaths were taken, and for the eight days that followed there was a reign of terror that overshadows altogether that other terror of Paris.

All statistics in relation to new discoveries are unreliable, but it is claimed that 30,000 Frenchmen were slaughtered. Certainly the estimate is exaggerated. Some escaped across the Windward Passage to settle Guantanamo on Cuba, some escaped to Tortuga; some, in the towns, survived. Yet the era of France in Haiti ended with the eighth day.

What followed is disintegration.

In 1794 the emancipation of the slaves of Haiti was officially announced. It was already in being. In 1801 François-Dominique Toussaint l'Ouverture, greatest of the figures who rose out of the confusion of campaign and massacre and the special internal struggle that ensued between the mulattoes and the blacks, declared himself perpetual Governor General. Driving the last French governor out, he invaded the Spanish half of the island, forced the capitulation of Santo Domingo, and freed the negroes of all Hispaniola. Napoleon sent a force of 25,000 men to restore control under the command of Victor-Emmanuel Leclerc, his brother-in-law, Toussaint was enticed treacherously on board a warship and taken to France to die in a fortress in the Jura. Where he had failed, fever took over the campaign. General Leclerc died and a great part of his men, and after extraordinary barbarities under Rochambeau the battered French army evacuated the island.

Immediately a third dictator arose. Dessalines, one of the leaders against Leclerc, had himself crowned as the Emperor Jacques. With Dessalines the simple alternation of racial conflict comes to an end. Before this, it had been white against brown, black against white, mulatto against black, and black against white again. Hereafter it is black against black with mulatto undertones.

Dessalines inherited a wrecked imperial economy. He replaced it with a tyranny and was appropriately assassinated two years later. The country divided in two. Henri Christophe set himself up as King of the North.

As the generations change there are different evaluations of the dark names of Haiti. Christophe, to one school, was a great patriot, a soldier of tremendous parts, a patron of the arts; to another, he is the archetype of black tyranny. On one of the highest points of the northern mountains, the Bonnet à l'Évêque, he built the great pseudo-medieval fortress that still towers there. It is the symbol of an incredible reign. Local history claims that it cost thirty thousand lives to build, that from its terraces Christophe marched whole platoons to death to establish discipline, that in it his 'madness' reached its pinnacle. Figures, again, mean nothing. His castle meant less. As a military work it was as pointless as the regime Christophe initiated. It had an exterior grandeur. It commanded nothing. It left behind it only a tradition of horror. After Christophe there is little more than a succession of aspirants to tyranny.

For a hundred years president followed president – literally at times, for fifteen were driven into exile, four were assassinated, one only retired of his own will at the end of his constitutional term. In 1915, following yet another revolution in which 200 political prisoners were bayoneted in the gaol at Port-au-Prince, American Marines took over control. For almost twenty years America enforced a resented stability.

Occupying forces are logically unpopular. Inevitably there were mistakes. But at the end of the occupation the Marines left Haiti peaceful – at least in the Central American sense of the word – economically sound – again in the Central American sense – and with a large programme of public works under way. There were even elections that were not disputed by the bomb and the gun. Despite occasional setbacks, enforced stability appeared likely to survive.

Always in the background there was, of course, the dark and tortured anti-religion of Voodoo. In the night the drums still beat across the Island of the Mountains, and in the *tonnelles* the hysterias of possession enacted their prehistoric mysteries as the old gods of Africa, inextricably mingled with Christian saints, descended and inhabited the bodies of the initiates.

How evil is Voodoo? No one perhaps knows precisely. Anything may flourish, it appears, under its communal hallucination. Probably the reported ritual murders, the occasional rumoured cannibalism, are in fact outside the cult, confined to the magic groups – the Mazanxa, the Mackanda or, above all, the Vlinbindingue. It is of little consequence. No one knows how many of the Houngans, the priests of the Voodoo cult, are members of the groups; no one knows how much they control in politics.

In 1957 Dr François Duvalier, a quiet-spoken country doctor who had been Minister of Health in the previous regime, was elected President in a peaceful election. There was, inevitably, a ritual challenge of corruption and a small display of force. It was suppressed. Duvalier was educated, intelligent, and had acquired a reputation as an earnest practitioner in an impoverished countryside. A new era, it was apparent, had opened.

Yet instantly the impalpable influences of Haiti began to operate. Within six years the quiet country doctor was accused of a tyranny comparable with the high line of his predecessors. The second election was a farce. To save himself the bother of another, in 1964 Duvalier declared himself President for life 'in response to public demonstrations'. Three years later he exiled his two daughters and a son-in-law for 'plotting against his life'. He was alleged to have immersed himself in the Voodoo cult.

Simultaneously the economic situation of the island deteriorated. Officially the *per capita* income of the Republic fell to $70 a year. Cynical American opinion declared it to be as low as $15. American aid had long since been cancelled, the valuable tourist trade had collapsed. Ninety per cent of the population was admitted to be illiterate, and under 'Papa Doc' the Tonton Macoute, the private army of the 'bogy-men', exerted all authority. In 1968, in an interview with *The Times*, the quiet country doctor claimed that 'not only as a duty but as an obligation' he had personally conducted the execution of nineteen army officers for plotting against State security.

The wheel had moved from Boukman to the twentieth century. It had remained in the same track. In the night, as one slips under sail through the Windward Passage, it is always possible to hear the throbbing of the Voodoo drums. Whether at the height of his Voodoo phase 'Papa Doc', as was reported, claimed immortality or not is perhaps immaterial. One night in March 1971 he died, leaving the succession to a son remarkable even by Haitian standards.

The drums throb on.

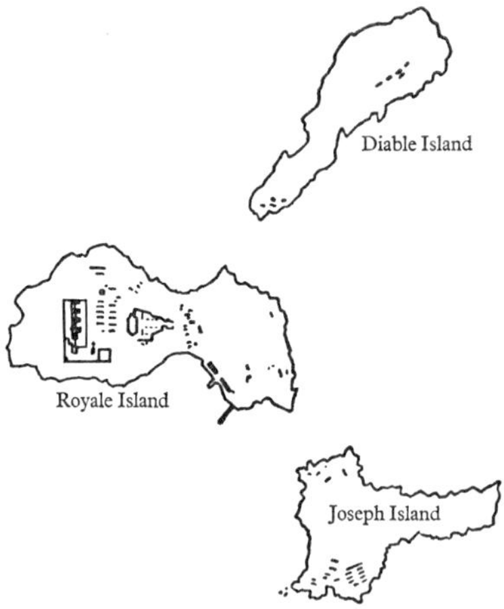

DEVIL'S ISLAND

Devil's Island is the most northerly of a triangle of insignificant islets 320 miles above the equator, twenty-seven miles north-west of Cayenne, and six miles off the mainland of South America.

Its discovery is undefined, but the group was named almost certainly when Daniel de La Ravardière landed at Cayenne in search of *el dorado* – the golden man – in 1604. Like others in that hopeless and romantic questing he did not find him, but the islands, with an unconsciously ironic prevision, were christened the Îles du Salut – the Islands of Salvation.

Île du Diable, as has been said, lies to the north – rocky, barren, with a sparse group of stunted palms, wholly inconsequential. Île Royale, the big island – it is itself only just over half a mile in length and less in breadth – lies at the inshore apex of the triangle, separated from Diable by a channel 200 yards in width. Île Saint-Joseph lies to the south-east. The group rises from a shallow patch never more than three fathoms in depth, which itself rises from the five-fathom shoal that fringes the South American coast in these latitudes. It is not a beautiful group. Its islands lack even the general blessing of the tropics – blue water. The sea around them is at its best a greenish yellow, frequently it is a turbid brown with the effluent of the Amazon, the main drain sewer of Brazil.

For a hundred years the area had no history, except of failure. Small expeditions, most of them originating in Rouen, attempted to establish settlements on the mainland; four of them collapsed in swift succession. Then in 1763 the Duc de Choiseul, the Pompadour's protégé, conceived a grandiose scheme of colonization. An astonishing fleet was assembled, 12,000 'volunteers' embarked and, flushed with enthusiasm by promises, descriptions and guarantees that would shock a present-day land shark, sailed for the earthly paradise.

Nothing had been made ready for them: neither shelter nor food nor even, in that place of torrential rain, drinkable water. There was no prophylaxis against fever, no medical help for the bizarre and innumerable tropical diseases. In the first months the volunteers died by scores. In the next by hundreds. At the end of the second year 11,000 were dead, 918 fever-stricken wrecks survived. The legend of the horror of Guiana was firmly established. The Duc de Choiseul continued in the exquisite sophistication of eighteenth century Paris, elegantly fostering the basic causes of the Revolution.

The Revolution spent itself, and Guiana came to the forefront again. Sentenced Royalists were exiled there. Six hundred were settled on the Sinnamarie River; 400 perished in a second tropical immolation. For the next half-century there were intermittent efforts at the most respectable type of colonization. They failed as miserably.

The way in which the aura of evil takes possession differs with every island. Easter Island is evil by the accident of fate, assisted only in its later history by the malignancy of man. Haiti is evil predominantly by reason of greed, racial arrogance – white *and* black – and only obscurely by fate operating through the hallucinatory medium of Voodoo. Devil's Island is evil by government decree.

In 1852 the prisons of France, with the combined results of the Industrial Revolution and the political rebellions of 1848, were filled to capacity. A prison building programme would have involved expenditures embarrassing to the third Napoleon's rash ambitions. An alternative was essential.

Exile has an ancient and more or less honourable history. It was a fundamental punishment of Greek law. In Rome it took the place under Augustus of interdiction, with a specific form of banishment: *deportatio in insulam*. Sardinia was one of the most dreaded of the islands of banishment, Gyarus another. Crete, Cyprus and Rhodes were considered by the cognoscenti as tolerable.

The President Napoleon, about to become the Emperor Napoleon III, had, however, a much later precedent, the precedent of Guiana. It is true that he was born too late to remember the exile of the Royalists – but the papers were available.

In 1852 a penal settlement was established at St Laurent, forty miles inland up the River Maroni, and rather more than eighty miles north-west of the Îles du Salut. A special settlement for incorrigibles was set up at Kourou on the coast and a conventional prison for the most hardened prisoners was built in due course on the Île Royale – *not* on the Île du Diable.

The settlement changed with the years. At first it was basically a place of political exile, long-term offenders of the Revolution of 1848 and in due course *communards* from 1871. Finally, it included recidivist criminals. It acquired early a brutal reputation compounded of harsh conditions for harsh prisoners and the normal abominations of climate and topography that had broken previous attempts at settlement. There was still no prophylaxis against the fevers of the tropics, no relief for the skin diseases and heat stroke, the insects and the damp.

It was unpopular in the underworld of France. There was occasional self-righteous criticism from outside. Britain had given up Botany Bay and all that that implied, even if the Indian Government was convinced that the Andamans were a satisfactory solution to its problems. France as a whole accepted that Guiana was a convenient point at which to sweep the dust under the carpet, and France traditionally was impervious to criticism. The settlement developed – progressed would be an unsuitable word.

Certainly it was a hell. The tropics anywhere in the 1850s were atrocious without ice and servants and European comforts. Like the rest of Central America, Guiana's fevers were both endemic and epidemic. Its skin diseases were those of all humid and over-heated areas. Its forests were picturesque from a ship but abomination to exist in. And the French exercise of legal process provided for it a steady supply of human fodder.

Its conditions, of course, resulted in a diminished expectation of life. Over the ninety-four years of its existence its total intake was 52,000 men, just over 550 a year, but though to that figure the dramatic addendum 'and few ever returned to their native land' is always made, the majority of the deportees to Guiana, in fact, lived on under sentence of perpetual exile. Simple arithmetic proves that the death rate of the colony never even remotely approached the disasters of Choiseul's 'volunteers' or even those of the Royalists of the Revolution.

It was, none the less, an abominable place. It was also, over the first half of its existence, never known as Devil's Island.

In 1894 Alfred Dreyfus, Captain in the French Army, was tried, found guilty, and sentenced as a traitor to France. The verdict condemned him to 'detention in a fortified area for life'.

Accident, intrigue, corruption, prejudice and anti-semitism played their part in one of the most dramatic judicial miscarriages of modern history. Three months after sentence Alfred Dreyfus was deported to Guiana.

The Île du Diable was not, before that date, used as a prison island; though a British Admiralty chart of 1870, incorporating French surveys, shows ten small buildings on its half-mile length. These buildings had no plan, there was no road, there were no encompassing walls. The Île du Diable was, in fact, the leper island of the area, providing isolation in the customary manner of the time.

Instructions from Paris decreed that Dreyfus was to be held in solitary confinement on Diable. In March 1895 he was disembarked on the Île Royale and accommodated at the main prison while a special hutment was built on the leper islet. It consisted of a room thirteen feet square with an antechamber for the guard about half that size. Five warders were allotted, surveillance was perpetual day and night, but the conditions were in no sense 'inhuman'.

> 'By day,' says Dreyfus in his moving account of his prison life, 'I was allowed to go about only in that part of the island comprised between the landing-place and the little valley where the lepers' camp had been, a space of about two hundred metres and utterly bare.'

In the first eighteen months of his exile, in reality, his circumstances were relatively tolerable. He had ample opportunity for exercise, he had the sea, and if a corrugated-iron roofed building in the tropics argued desperate heat, he had at least the frequent solace of the sea breezes. His diary, however, is a chronicle of endless grievance and complaint. He had to cook his own food and he was unable to acquire even the rudiments of the art. He was subject to all the endless petty tyrannies of prison life: his mail was delayed, books failed to reach him, he had nothing with which to wash his exiguous cooking utensils. Surveillance worked constantly on his nerves, the surveillance light at night was a torture, he suffered intermittently from fever, he was convinced that he had 'congestion of the brain'.

In terms of the Guiana penal settlement as a whole and of the general French attitude to convicted persons, his treatment was, none the less, comparatively unremarkable; a convicted traitor might even have considered it endurable.

The essential point about Alfred Dreyfus was that he was innocent. His irrational complaints against the detail of his imprisonment, his indignation at the affronts to his honour as an officer, the alternations of hope and despair over the delayed arrival of his letters amount to neurosis. The evidence of his own account suggests that he was neurotic even before his trial, certainly immediately after it. On Devil's Island in the second stage of his imprisonment his neurosis became pathological.

The second stage was brought on not because of local inhumanities but by the injudicious action of his brother Mathieu who, fearing that Dreyfus was in process of being forgotten, planted a story, in of all newspapers the *South Wales Argus*, that he had escaped. In September of 1896 the London *Daily Chronicle* reprinted the story. The primary result of this attempt was an order from Paris for a double palisade to be built round Dreyfus's hutment and for Dreyfus himself to be put in irons each night until it was complete. The palisade shut out the sea breeze, Dreyfus's walks were restricted to the narrow lane between its walls; his imprisonment became for the first time absolute.

As a name Devil's Island had obtained currency before this; in the middle of September it achieved official cognizance when General Ganse, an Assistant Chief of Staff, answered a protest with the question: 'Why do you make a point of Dreyfus's leaving the Devil's Island? The case can't be reopened.'

General Ganse was wrong. It was at the point of reaching the summit of its drama. An astonishing farrago of accusation and counter-accusation, press interference, manipulation of witnesses, duels, suicides, and government scandals began its inevitable course. It developed beyond even the capacity of Zola, one of its major protagonists, to invent. From now on through all the infinite complexities the name Devil's Island appears in every account, in every newspaper description, in every discussion. It acquired a black romance; it was inflammatory, it was evocative, it was inexhaustibly damaging to the brutally preserved respectability of the French High Command, and it incorporated into itself all the extraneous evils of the penal settlement as a whole and all the horrors of the Royalist exiles and the volunteers of the Duc de Choiseul. In the end this scabrous little island with its single group of discouraged palm trees and its barren shank became the living symbol of the horror of Guiana – and of the perversion of Justice.

In the upshot Dreyfus was taken back to France, tried again, given a free pardon, exonerated, restored to the French Army.

The penal settlement went on. Its pattern now was a pattern of escapes from Devil's Island. They were, actually, escapes from St Laurent on the mainland where the convicts could obtain boats or canoes, drop down the Maroni river and work up the coast of the Guianas to Trinidad, but the name found official recognition even in the Judicial Committee of the British Privy Council. A case reported in the London *Times* and giving the Committee's judgement appears under the heading 'Fugitives from Devil's Island', but after Dreyfus there was scarcely ever anyone to escape from there; more certainly, nobody did.

Somewhere about 1924 France began to consider the abandonment of the settlement as a whole, not altogether on moral grounds but because in the slump that followed World War I it was discovered that the penal settlement was costing France a quarter of a million sterling a year. Nothing, however, happened. In 1935 a Consultative Committee of the Chamber of Deputies declared formally that it was no longer wanted. Nothing happened. In 1938 its abolition was decreed. Nothing happened. It was still there through World War II, when warders were alleged to have assisted in mass escapes. Not until 1944 was it positively abolished, and even then in 1946 and finally in June 1947 men were still 'returning from Devil's Island'. In 1947 the prison on the Île Royale itself was at last abandoned, the punishment camp at Kourou was closed, and St Laurent decayed into a tropical slum. And in Paris 'Rene Boismoreau, sixty years old, bespectacled, toothless', told a reporter of the London *Daily Express*: 'I could get a drink for a penny in Guiana, but here they want sevenpence halfpenny for a small glass of wine. I won't stay unless prices come down.'

Under the tropical sun and in the wet winds the prison on the Île Royale crumbled. The corrugated-iron church with its frescoes of the crucifixion by Georges Frederic La Grange – one of the most distinguished forgers of his time – survived as a rusted ruin. The administrative buildings were roofless, the cell blocks shattered. A correspondent of the London *Sunday Times* in 1962 found on a wall a single inscription linking an age of past horror with the present. It said in large letters: *Vivent les Beatles!*

One year later General de Gaulle decided that it should be converted into a missile-tracking station.

At the moment of writing it is, under another government, 'for sale'.

The City Islands

PARIS

Paris is an island.

It owes its origins to that fact. It owes its restorations after repeated disaster to the same essential reality. That it lies not in the sea but in a river does not in any way affect its primary characteristics. Throughout its existence it has had always the qualities of an offshore island strategically placed between areas of desirability – a Crete between east and west, a Malta between the Mediterraneans, a Singapore on the divide between the Pacific and the Indian Oceans.

It is not more than three-quarters of a mile in length and it was smaller still in the time of its foundation, but with rare intervals it has held – because of those qualities – a dominant place in the pattern of Western Europe for 2,000 years.

The Seine is a lazy river. After it receives the Marne, greatest of its northern tributaries, on the outskirts of the modern city, it meanders exquisitely and irresponsibly between the city and the sea. The air line distance from its source north-west of Dijon to the Baie de Seine is almost precisely 250 miles – the course the river takes 482. Inevitably it spawns islands. Between the junction with the Marne to the east of the city limits and the junction with the Oise to the west, alone, there are a score of them – long, narrow slivers of land that began as 'middle grounds', shoals that were built up with the deposition of mud and sand and shingle in the winter freshets, mud-banks that were swept together by the contemptuous swirl of the current, washed altogether away by the next flood, or left alone to acquire solidity. Weed and reeds and grass bound them; they graduated to become dry ground with willow trees and alder. Some of them are delicately beautiful. One of them is the Île de la Cité.

In the time of the Gauls it lay within the westernmost fringes of the territory of the Sequani, whose heartland streched up the Seine and across the divide to the Saône. The Sequani granted it to the Parisii, though it may have been inhabited fitfully by a La Tène people a couple of centuries before. The Parisii were described by contemporary account as 'a small tribe', but they were capable, none the less, of sending 8,000 fighting men to the army of the Gauls at the last battle for Alesia.

It was a trading tribe; one offshoot had crossed to Britain in the great Gallic migrations, to settle in Yorkshire above the Humber – the Lady's Tomb at Arras on the Yorkshire wolds beyond Beverley is identical with chariot burials found close to Lutetia in the valley of the Seine. This island they sought for purposes of trade. They called it Lugoticia or Lutentia or Lucotecia – there are many variants. Rome softened it to Lutetia.

In the earliest beginnings of Gaul the constant and restless motion of the tribes wore a pattern of tracks across the country. One of these at least, more probably a conjunction of two, crossed the Seine at the fords of Lutetia. These were by no means the only or even the lowest fords of the Seine, but here the river ran wider than it ran to the west between the high ground of Sèvres and of Bologne and was therefore more shallow, and the island broke the crossing into two. One of the tracks ran south-east to Autun and the valley of the Rhône, and so to Eastern Provence. One ran south to Orleans, crossed the Loire, and appears to have passed in time through Periguex to Narbonne and the Mediterranean trade. North of the Seine a single track forked a few miles beyond the ford, westward to Rheims and north-north-west to Amiens and the Channel coast.

The importance of Lutetia, however, is not the simple importance of a river crossing. The river itself was, even before the Romans came, an important artery of trade. By the settlement of the Parisii, Lutetia was established as the nodal point in the development of the incipient commerce of Central Gaul.

It has, however, no place in the first campaigns of the Roman conquest. Caesar mentions it only after his return from the second invasion of Britain. In 54 B.C. he brought his main force back to Amiens to a disastrous failure of the harvest. Hastily preparing a new plan for winter quarters to make the best of the limited grain reserves, he called a council of the tribes. The new plan spread the legions in a wide separation over Northern Gaul.

Like other conquerors after him, Caesar underestimated the potentials of the spirit of independence in a conquered people. Early in the winter the Carnutes, and after them the Eburones, grasped at the opportunity to attack a separated legion. A series of brisk winter campaigns developed, which ended with a sharp pre-emptive attack on the Nervii. By the spring the situation was inflammable. Caesar summoned a second *Concilium Galliae* at Amiens. Three tribes failed to attend. The failure was tantamount to a declaration of war but, before challenging it, Caesar called a third council – at Lutetia. For the first time the name appears in written history.

The time for councils was past. Intermittent fighting continued through the whole of the summer and the greater part of the second winter. It was bitter in the spring. With the first good campaigning weather, Caesar split his forces and sent Titus Labienus, with four legions and a strong force of cavalry, against the Senones and the Parisii, while he himself attacked into the country of the Arverni.

Labienus advanced rapidly from Sens into the valley of the Seine and made contact with the veteran Camulogenus. Camulogenus had entrenched himself strongly with a mixed Gaulish army in the marshes to the south of the Seine. Unable to get at him across uncertain and treacherous ground, Labienus struck upriver again, searching for a crossing. The bridges at Lutetia were strongly held and he attacked at the smaller town of Metlosedum, itself also upon an island. Metlosedum burnt its bridges, but Labienus, with frenetic energy, seized fifty river craft – the number is an index of the existing scale of the river traffic – constructed a floating bridge, crossed, sacked Metlosedum, and transferred his army to the northern bank of the Seine. Sweeping down the river, he probed at Lutetia. They burned the bridges in his face and the town with them, and he camped somewhere about the site of the Madeleine.

Despite his energy, his position was precarious, for the Northern tribes were rising. Labienus developed an elaborate deception plan to persuade Camulogenus that he was withdrawing, flung his army by night across the Seine in a masterly amphibious operation instead, and at dawn had three legions in line. Camulogenus, rashly permitting himself to be drawn from his marsh defences, was destroyed.

Disregarding the ruins of Lutetia, Labienus headed for Sens again, and the southern battle.

Lutetia was rebuilt. First, almost certainly, the local people restored the sunken wicker houses of the trading settlement, for Rome at that perilous moment had no troops for reconstruction – Alesia was still defiant. In time, and a welter of slaughter, Alesia fell, and the spirit of independence fell with it. The *pax Romana* was imposed across Gaul.

Lutetia is described early in the peace as 'fortified'. The archaeological evidence is inconclusive. Had there been even a cohort fort, the plan should certainly have been discernible. It is not. At most it may have had a bridge guard. Yet its strategic importance was clear in the light of Labienus's problems. Its commercial importance was plain enough before.

In a brief time it grew and overflowed the narrow area of its island. Across the river to the south there was excellent land for a spreading town on the rising slope of the Montagne Sainte-Geneviève. Beginning perhaps like a *vicus* at the gates of a legionary camp, the left bank town expanded with sea trade. Under the foundations of the apse of Notre Dame, where the temple quarter of the Roman city lay, excavations uncovered a Roman altar. Its inscription declares that it was set up by the *Nautae Parisiorum* – the shippers of Paris.

There is an engaging and admirable *maquette* of the city at this period by Bordier, which shows the island weighted down by close-packed buildings. The temple stands at the north at the head of an oblong colonnaded forum. Barracks blocks, storehouses, dwellings fill every foot between the forum and the water's edge. To the south stands the palace of the governor. In the downstream spear-point of the island is a garden. The bridgeheads are walled and defended, and the defences are repeated on the island. It looks across the southern arm of the Seine to the straight streets of the left bank town, the vast pile of the Northern Thermae (the ruins of which are still immense in the Musée de Cluny), the arena to the south-east, the theatre to the south.

In the third century the *pax Romana* failed; the left bank was burned.

A new evaluation of the security of Lutetia was swiftly made. For the first time a wall was built. It followed the whole coast of the island. They used the ruins of the left bank as a quarry for its material – the seats of the arena, the ashlar of the theatre. It was turretless but, with the broad waters of the river at its feet, it was astonishingly strong. The island quality was still to the Romans the primary consideration.

It was in this new-walled town that Julian of Byzantium, the apostate who returned to the older gods, was declared emperor by the men of the rebellious legions. It was in this town that he wrote, with a humanity rare in emperors:

> My dear Lutetia, built in the midst of the river upon a little island, wooden bridges joining it to the two banks. This river rarely fills too full, rarely runs too low, as it is in winter, so it remains in summer. The water is excellent for drinking, the climate soft and gentle, perhaps from being near the sea; the vines are good in quality and very numerous.

Dear Lutetia . . . her glory under a Roman emperor was very brief. Julian went east to Byzantium, and politically in the later years of the Haut-Empire Paris was inferior to Sens. Religiously also – Sens had an archbishop.

Paris, on the other hand, had Sainte Geneviève, and it was Geneviève at the age of twenty-nine (she had been *religieuse* since the age of seven) who made one of the great strategic appreciations of Roman Gaul. In the face of the stupendous threat of Attila and the Huns, Geneviève persuaded the people of Paris to 'stay in their island'. They did. The Hun flood poured south, evading the complex problems of an attack upon an island fortress. Paris was saved.

Very properly, Geneviève was canonized, but Rome in due course fell and Paris passed to the Franks. Clovis, the conqueror, was however, not a destroyer. Paris became the capital again – the first capital of the first 'France'.

For two and a half centuries the Merovingians held Paris – most of the time in an intricate mosaic of civil war, family strife, decadence and degeneration. Charles Martel, Pepin the Short, and finally Charlemagne, re-established power, but in the process Paris lost her status. Charlemagne moved the capital to Aix-la-Chapelle.

Strong kings tend to leave weak successors. Weakness in France coincided with strength in the North countries. In the year 845 Horik, King of Denmark, evaluating its situation as an island, sent Ragnar Lodbrok with a fleet of long ships up the Seine to capture Paris. It was the beginning of seventy fantastic years in which for a large part of the time the Danes, based on the island of Jeufosse near Vernon and a little over eighty miles below Paris, controlled the valley of the Seine and strangled the trade of the capital.

The classic campaign of island against island in the midst of the rich, rolling valley of the river lasted for almost seven years. The problem of the Northmen lasted until, largely to free the capital and its strangled trade after the ravages of the 'Great Army' of the Vikings, Charles the Simple granted the coastal areas to Rollo, King of the Northmen. Thereby he set the seed of the Norman invasion of Britain under William and of the return invasions that were to plague Paris for the next 800 years.

Blockaded, beseiged, burnt, overrun from without, broken by revolution from within, Paris moved through the centuries, always growing. The successive lines of fortifications are like ripples spreading out from the Île de la Cité as ripples spread from a stone in a silent pool. Romanesque Paris developed and dissolved. Gothic Paris aspired and came back to earth. Renaissance Paris created itself in the austerities of geometry and overwhelmed itself in the vulgarities of Baroque. And all the while the Île de la Cité endured, calm in recurrent storms.

About it today is the maelstrom: rocks, reefs, ridges of the splendours of antiquity rise out of the banalities of the nineteenth and the froth of the twentieth centuries implacably enmeshed in the pulsating pattern of the traffic. The Louvre and the Tuileries and the Place du Carrousel. The University and the Musée de Cluny, and the little green corpse in the purple tomb of the Invalides. The third and more improbable Napoleon and the Baron Haussmann. The detritus of a score of international exhibitions, each uglier than the last. The upper middle class hauteur of the church of the Sacré Coeur. All of it neatly parcelled for the tourist trade.

At night this other Paris roars like the trade wind surf on the reef of a Pacific atoll, but the Île de la Cité is silent. Its downstream point juts into the fleeting water like the prow of a ship. It is moored to the noisy banks by ten bridges as if the rest of the city were afraid that it might slip away.

No one moves – except across the bridges. The garden at the spearpoint where the Roman garden grew is utterly quiet, except when the voyeur beam of the searchlight of a tourist boat fingers it, searching for courting couples. The Palais de Justice is wholly dark, and even the houses in the Place Dauphine show few lit windows. The Sainte

Chapelle is still and secret, and the vast bulk of Notre Dame is only a floodlit silhouette against the neon of the east.

Still or waking, it remains the heart of France – the Judgement, the Prison and the Church. In the dawn the Sainte Chapelle wakes to the ultimate splendour of coloured light, the *gardiens* open the doors of justice, and high on the towers of Notre Dame the *chimères* stare out over the other city with an infinite weary cynicism.

Paris is still an island in the Seine.

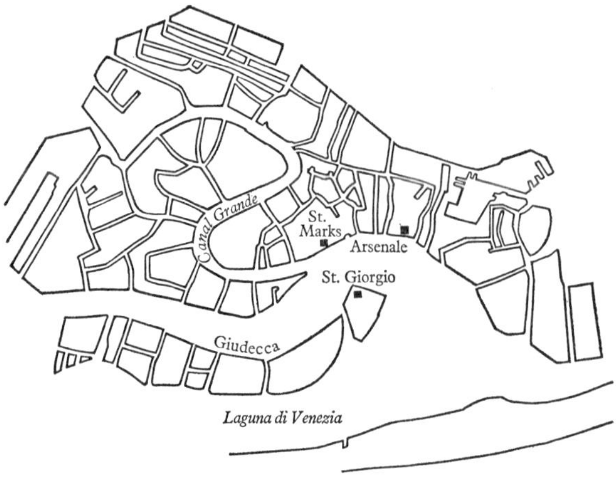

VENICE

Venice is not an island

Venice is an archipelago – a complex of 117 islets: separated by two deep-water channels: sub-divided by 150 canals: crossed by 300 bridges: bordered with palaces.

Statistics are inappropriate to Venice, meaningless in the presence of the exquisite. The single, solitary necessary fact about Venice is that it rests on mud.

Its shape presents no problems. If you fly to the southward, you will find a score of sites for another Venice in the Laguna Morta. The islands north of Torcello in the Palude della Rosa show what it was before man came. If you fly above it, high enough above so that the reddened umber of the roof tiles merges with the low-tide shallows, you will see precisely how it was formed – the river channels converging towards the gap in the *lidi* at Castel Sant'Andrea, the subsidiary canals draining the sandbanks as the tidal creeks drain the shoals of half the river estuaries of the world.

The Adriatic current swings round the head-waters of that sea from east to west. Over the geologic ages it has modelled the sand and mud of the mountain streams into the chains of the narrow *lidi* from Grado to the Valli di Commáchio. Inside the breakwater that they formed, the rivers built their own shallows where the incoming tide checked the outward flow, and the sand and mud sank slowly to the bottom.

Venice was not discovered. It existed. Long before Attila thrust that way in the middle of the fifth century it was known certainly to the fishermen of the Veneto, inevitably to the people of the early cities: Aquileia and Concordia in the north; Altinum, Padua and Vicenza.

It was known and unregarded. Some of the islands in the lagoon had fishermen's huts on them, there were no built houses; archaeology has discovered less on these empty sandbanks than under any other town in history. But it *was* known – and because it was known, it offered refuge. When Attila and the Huns burnt Aquileia and stormed through the cities of the Veneto, their people fled to the lagoons. When Attila went north again, they moved slowly back to their ravished homes.

There is no single moment of stability in the establishment of the complex of the lagoon towns. All through the turmoil that followed the collapse of Rome, armies and races used the broad plain of the Po as a looting ground. As the people of the towns moved in and out of refuge, the sheltered places of the lagoons grew into towns themselves: Grado and Bibione and Caorle, Heraclea, Jesolo, Torcello and Murano, Malamocco, Poveglia, Chioggia, Sottomarina – and the Rialto. They were distant enough, and in time they were strong enough, to defy the Lombards.

They set up a government of the lagoon. Tribunes were elected from each island. In the immemorial pattern of Italy, they set up simultaneously the traditional internecine rivalries. But in 697 the twelve towns, brought together by common fear, elected a common leader, a *dux* – the word corrupted into doge – Carlo Lucio Anafesto.

There was a Doge, there was still no Venice.

Power grew. Despite the inherent rivalries, the lagoon towns as a group moved nearer and nearer to statehood. Finally Pepin, son of Charlemagne and King of Italy, having shattered the Lombard kingdom, made up his mind to break this second incipient threat.

Capturing Chioggia at the southern entrance to the lagoon, Pepin began a slow advance up the long sandbank of Pellestrina aiming at Malamocco, the compromise capital of the twelve towns. The imminence of danger united even the factions. In A.D. 812 the government, by unanimous decision, withdrew to the central group of islets and the very small town already built on the Rialto. There, safe in the intricate maze of the waterways, superbly protected by the deep mud that lies under the eighteen-inch water depth that covers most of the lagoon, they stood in defiance.

Pepin retreated, Venice was born.

Little enough is known of the early town of the Rialto. There was a church where St Mark's stands, built in honour of St Theodore, patron saint of the little town – and his crocodile. There was a green field in front of it, according to Galliciolli:

> . . . partly covered by turf and planted with a few trees; and on account of its pleasant aspect called Brollo or Broglio, that is to say, Garden. A canal passed through it over which was built the bridge of the Malpassi.

There were other churches. San Giacomo di Rialto is claimed to be the oldest foundation in Venice. San Giorgio Maggiore, on its islet, was rebuilt upon tenth century work. The houses were probably single-storied, enclosing a courtyard, sometimes with a loggia on the roof. The official buildings must have been small like the tiny palaces of Torcello, the bridges infrequent.

Agnello Participazio, the first Doge actually elected on the Rialto, began with immense energy the work of expanding the town for the requirements of an ambitious independence. The islets were reinforced with sand, the banks strengthened. Wooden bridges were built over creeks that had been deepened into canals. Pile work was perfected. And at length, at the corner of the Bridge of Straw and the Rio di Palazzo, he began the construction of a ducal palace and with it, or a little later, the reconstruction of the small church of San Theodore as its chapel.

In 628 the body of St Mark, thoughtfully purloined from Alexandria, reached Venice. St Theodore and his crocodile were ousted; St Mark, with his winged lion, became the patron saint of Venice. Apart from the market importance of his body in an age when the trade in relics had achieved dizzy heights of inflation, St Mark had a particular connection. It was said that, bound for Aquileia and fighting against a gale off the *lidi*, he saw 'near the islands of the Rialto' a vision of an angel.

Everything depends on what is meant by 'near'. The islands of the Rialto were then, and are now, invisible over the ridge of the Lido. In Mark's time it is possible that they lacked even the importance of a name. But in the new chapel of the palace the body was interred with immense solemnity.

The great era of expansion had begun.

Like Paris, the city of Venice was established for trade. Like Paris, it stood between areas of desirability – the eager markets of Europe on the one hand, the illimitable riches of the East on the other. Like Paris, Venice flourished.

As the new capital rose from the water, therefore, it attracted the attention of the Dalmatian pirates, who hung like hounds on the flanks of the merchant convoys and harassed the *lidi*. In the year 1000 the Doge, Pietro Orseolo II, sallied from the lagoon, captured Curzola on the Dalmatian coast, and razed the pirate stronghold at Lagosta. A little after, and farther down the Adriatic against the Albanian coast, they checked those other pirates – the Normans. With the victories, Venice secured the mastery of the Adriatic; the way to the world was opened. Inside the lagoon, and ignoring St Mark, Venice conceived a ceremony as pagan as anything in medieval history. Every year on the Feast of the Ascension, Venice married herself to the sea and a ring was dropped into the Ocean in an older propitiation than Christianity.

There are no pictures and few records of this Venice. Yet it had by now acquired, clearly, a remarkable dignity. Cassiodorus had written to the Emperor Theodoric a description of the earlier settlements: 'like seabirds' nests over the face of the water'. His ornithology was weak but his description was sound enough. This new capital, however, was a major town already, and already it had suffered major disasters. The ducal palace was destroyed in the fires of insurrection against the Doge Candiano; the ducal chapel that had been constructed to hold the body of St Mark disappeared in the same fires – and the body of St Mark vanished with it. Orseolo rebuilt the first two; and the campanile of St Mark rose at the corner of the piazza which had replaced the old green meadow – to perhaps two-thirds of its present height. The body of St Mark was miraculously re-discovered.

Otho the Great's chronicler says of the emperor's visit to Orseolo: 'The Emperor beheld carefully all the beauty of the palace.' Venice had advanced far past the first hamlets of the refuge. The merchants of the early trade had been translated into merchant princes, their leaders walked with kings. Already the great homes along the curves of the Grand Canal and fronting the little squares were themselves palaces, the wooden bridges were being replaced in stone, everywhere churches were rising. From the envious rooftops of Torcello to the north the city must have lifted along the horizon like a rampart.

Venice was sufficiently powerful now to play agent to the Crusades. She had ships enough and she controlled the routes to the Holy Land,

and from the first three Crusades she secured vast trading rights in the Levant and the destruction of the rivalry of Tyre.

The long quarrel with Genoa rose out of this. Quarrels with Constantinople grew simultaneously, but still the power of Venice increased, and in the Fourth Crusade it was to her alone that the leaders applied for transport. For 85,000 silver marks of Cologne and *half of all the conquests* Venice agreed to furnish transport for 9,000 knights, 20,000 foot soldiers, 4,500 horses, and provisions for a year. The logistics are stupendous; they are the measure of the growth of 400 years of the most brilliant, the shrewdest, the most ruthless trading empire that the world had ever seen.

The primary defect of the Crusades was that, unlike Venice, they lacked business management. They were organized on inspiration and run by religious enthusiasm – neither of which provided adequate substitute for administrative acumen. With the Fourth Crusade the money was not ready at the departure date, and the Venetians toughly refused to sail. Dandolo was Doge. With consummate skill he persuaded the leaders of the Holy Army to agree to an attack on Zara and Dalmatia (which had hived off from the Republic) before the proper business of the rescue of the Holy Places. It was to be considered as a *quid pro quo* for a cash postponement. Zara fell, and Dandolo, manipulating for a second time the scandalously malleable leadership, changed the direction of the Crusade altogether from the Holy Land to the great Christian Metropolis of Constantinople, in order to place a Venetian puppet on the throne of Byzantium.

Constantinople resisted, fell, and was pillaged. Venice claimed 'a half and a quarter of the Roman Empire . . .' and got it. The Cyclades, the Sporades, the eastern coast and islands of the Adriatic, Propontis, the Uxine and Thessaly were the profit of this most unscrupulous deal in the history of medieval Christendom. With them went control of the trade routes to the eastern basin of the Mediterranean, to Syria and Turkey, to the Black Sea and the Sea of Azov. Venice had made of her islets a European Power.

The Holy Places remained in the hands of the infidel.

With the new prosperity began the second of the great explosions of architecture in Venice. At the heart of things the Pala d'Oro, the golden retable of the High Altar of St Mark's, already magnificent, was made superb with the jewels of Byzantium. The Treasury was filled with its loot. The work on the glorious mosaics of the domes was accelerated, the great golden horses of Alexander were lifted in an architectural extravaganza to the roof of the fantastic atrium. Everywhere new churches were built – San Giacomo dall'Orio, Sant'Elena, Sant'Aponal – others that have since been rebuilt. By 1246 the Dominicans felt themselves firmly enough entrenched in the new prosperity to start San Zanipolo – the Pantheon of the Doges. Four years later the Franciscans, hot on their heels in the spirit of emulation, began the magnificence of the Frari.

Venice was almost Venice; not perhaps yet so tall, though many of the earliest of the existing palaces were built or building. Its reflections were not yet so long in the still surface, but its foundations went down 'steep-to' into the tide that is its salvation, the three-foot tide that is enough to run the diurnal stream through all but the dead-end canals, that flushes out most of Venice twice in the twenty-four hours and keeps it as alive as does the breathing of the human lung.

The palace of the Doges was rebuilding. It was always rebuilding – fire upon fire upon fire. And there was of course war with Genoa: there was invariably war with Genoa now, the jealousies over the eastern triumphs were not to be silenced. It was in the end a near run thing. Like the Dalmatians, the Genoese harassed the convoys. Like Pepin before them, they captured Chioggia and threatened the lagoon. Like the Venetians of Pepin's time, the city defied them. But it did more: it waited until the Genoese fleet was triumphantly in the harbour of Chioggia and it blocked the channel to the sea and held the block. The Genoese surrendered – Venice on its islands was safe again.

Already now it was aiming at land power. Forty years before Chioggia it had made its first venture in the March of Treviso. Now, in campaign after campaign Venice added villages, towns, cities. The campaigns went on into the fifteenth century. The great condottiere, Bartholomeo Colleoni, master-minded those of the middle of the century and was rewarded with that simple, splendid statue by Verrocchio on the Campo San Zanipolo (though not in the Piazza of St Mark where he had wanted it) that is one of the masterpieces of memorial. Venice spread to the gates of Milan, overlord of the plain of the River Po. By 1492, as Columbus entered the New World, it was at the zenith of its power.

Six years later it received its mortal wound. Very quietly, two whole oceans away, a little fleet of four ships slipped into Calicut on the coast of India, and Vasco da Gama broke the monopoly of the Eastern trade, the plinth of the Venetian economy.

Dying, for cities, is a slow process except by fire, earthquake or battle. Venice, for perhaps a hundred years, scarcely felt her wound. The third of her great architectural explosions began almost before the breaming faggots were piled on the beach of the Tagus for the return of da Gama's ships. The Scuola di San Rocco was designed. The Library of St Mark, one of the sumptuous buildings of Italy, was built by Sansovino across the Piazetta from the Palace. The eastern façade of the Palace itself above the Bridge of Sighs was constructed. Over the width of the harbour Palladio built San Giorgio Maggiore. Across the Giudecca he put up the Redentore in thanksgiving for the ending of the plague. On the far side of the Grand Canal Longhena built Santa Maria della Salute.

These are the six great public buildings of the new style, the buildings that figure in every picture of the heart of Venice, the buildings that, coupled with the best of the new palaces, are the master-work of the Venetian Renaissance.

Ruskin loathed them.

> Venice, once the most religious, was in her fall the most corrupt, of European States; in her strength the centre of the pure currents of Christian architecture, she is in her decline the source of the Renaissance.

It was Ruskin's unshakeable conviction, endlessly expressed in *The Stones of Venice*, that the fall of Venice began with the death of the Doge Carlo Zeno in 1418. Not all historians will agree. Certainly Venice in the succeeding century achieved depths of treachery – towards herself and towards Christian Europe – that were not to be matched afterward, but she also achieved triumphs. The process of decadence was slow and for a long time painless. The deliquescence of power was made supportable by the accumulated wealth of five hundred years. The decline of civic morality was made supportable by art. Ruskin maintains censoriously that the Venetians fell '. . . from pride to infidelity, from infidelity to the unscrupulous pursuit of pleasure'.

On the accession of the Doge Foscari in 1423, it is true, 'the city kept festival for a whole year'. But the process was intermittent. Venice was still strong enough a century and a half later to lead Europe – 111 of the 243 ships of the allied fleet were hers – at the decisive Battle of Lepanto, the battle that broke the sea-power of the Turk. Whatever the black involution of her internecine rivalries, the graph of her fall is interrupted by peaks of brilliance. Its descent is profligate, but it is illuminated by an endless succession of the great names of the arts: Giorgione and Bellini, Titian and Veronese and Tintoretto. It was even in decline a well-spring of beauty.

The real problem of the Renaissance in Venice is not its part in the fading revels but its intrinsic inhumanity. San Giorgio Maggiore is not so much a church as an essay in mathematical perfection, as intellectual as an exercise in Euclid – and as sterile. The mannerist complexity of its façade is matched only by the soulless chill of its interior. Yet Ruskin's commination is not to be upheld. San Giorgio is as essential to the city-scape of Venice as is the Redentore or Longhena's over-generous dome for Santa Maria della Salute. It would be as impossible to imagine the exit from the Grand Canal without these as the Canal itself without the Palazzo Grimani or the Camerlenghi or, for that matter, the Bridge of the Rialto.

Venice is an intricate, interlocking composite of three great architectures – the Byzantine, the Gothic and the Renaissance – and the special quality of Venice is that time and the reflections in the water have overlaid their enmities.

Ruskin, because of the violence of his love affair with the Gothic, could see, in the end, only decadence and decay. He saw the city

> . . . in the final period of her decline: a ghost upon the sands of the sea, so weak – so quiet – that we might well doubt, as we watched her faint reflection, which was the City and which the shadow.

A substantial ghost!

A century after Ruskin wrote that, Venice is bustling and triumphant. You can read it for yourself by way of Canaletto's brilliant canvas of the beginning of the ceremony of the *Sposalizio del Mar* – the Marriage to the Sea. The foreground is all gondola, black in accordance with the sumptuary laws, scattered at first but closing to a jostling shoal against the Schiavoni. Eight boats among them, in a splendour of gilding, counterpoint the fantastic golden opulence of *Il Bucintoro*, the barge of the Doge. The immense composition pulsates with excitement, with a passion for pageantry, with a worship of colour – and as a background to it all is the grave gaiety of the city itself: the rose and silver of the Palace of the Doges, the gigantic aspiration of the Campanile of St Mark, the columns of the Saints, the umber of the Palazzo Reale, the blue of the entrance to the Grand Canal, the mathematical perfection of Santa Maria.

It was painted halfway through the Long Carnival. It is instinct with the triple loves of luxury, of beauty, and of art. But this is not its miracle. The miracle is that after two centuries, and three world wars, Canaletto could paint it again today. He would needs do without the *Bucintoro*, he would have to include the vaporetti and the motoscafi, but the Venice that he painted lives.

She has fought for her survival in five battles that match the greatest in her past. Against the Railway Age, she won – and the railway stopped at the farthest fringe of her islets. Against the Industrial Revolution, she won – and her own industry was exiled to the mainland, while the necessary acceptance of the vaporetto was hedged about with safeguards. Against the Automobile, she won – and traffic, the destroyer of ancient cities, is penned to the Piazzale Roma. Against Electricity, she won – she is free even of the infinite banalities of neon except for that restrained blue sign next to the Torre dell'Orologio that says, with the unctuous insincerity of a Memphis Funeral Home, 'Wagons Lits/Cook'. And against the last enemy of all, the Air, she won – and the airfield is over against the Palude del Monte, and those who come to it at 600 miles an hour must cross to her at last, slowly, by water.

She has not, inevitably, emerged from the battles unscathed. Venice, like all the cities of this age, suffers from pollutions. Marghera, the area of the mainland to which industry was banished, drifts a corroding smog over her at intervals. Compared with the industrial cities of the Continent, it is, it is true, a minor imposition, but statues, churches and the façades of the palaces have developed a degree of 'marble cancer'. For all the regulations, the vaporetti and now the motoscafi still add a little each year to the undermining of foundations. The deep-sea ships passing up the Canale della Guidecca assist. The water-table in the soil below Venice has been tapped for industrial purposes and has added its contribution of depression to the rise in the level of the Adriatic that stems from a source as distant as the melting of the icecaps. 350 of the 450 surviving palaces of Venice are admittedly claimed to be in need of repair. But some of them have needed repair since the middle of the Long Carnival, and some at least of the surface erosion is due to domestic smoke, and the rise caused by the melting icecaps is barely .055 of an inch in a year. Venice has experienced less of the agonies of age than any other of the ancient cities of the world.

Venice is the city which has defied the twentieth century.

In the sunrise she is remote and exquisite. At noon she blazes. In the full of the moon she enshrines romance – suitably commercialized! She is most gloriously alive. She is still, in the reflecting water – *la Serenissima.*

MANHATTAN

It is a moot point whether Manhattan Island was ever discovered – in, that is, the accepted meaning of the term.

Giovanni da Verrazano, the Florentine, thrusting cautiously into the Upper Bay and 'all of a sudden as is wont to happen to navigators' baulked by a shift of the wind, certainly sighted a very large river: equally certainly he was unable in 1524 to explore it.

Éstevan Gómez, the Portuguese, following him less than twelve months later, paid enough attention to it to name it Rio São Antonio, and no more.

Neither of them recognized the island character of Manhattan. The Sea of Verrazano, as the Upper Bay came to be called, was known to be surrounded by a fragmentation of islands; but to the discoverers it was the river that was important, the river that was swiftly and with a highly unjustifiable optimism assumed to be a possible route to the Indies.

The area was visited by a series of ships in the course of the sixteenth century: anonymous explorers, adventurous traders, even perhaps fishermen, for the Basques and the cod-fishers of Brittany had discovered the Grand Banks close astern of John Cabot. A French map of 1570 shows the Bay in remarkable detail. But the nature of Manhattan was not yet explicit.

In September 1609 Henry Hudson anchored in the Lower Bay.

> Our boate went on land with our net to fish and caught ten great mullets of a foote and a halfe long a Peece, and a ray as great as foure men could hale into the ship.

The Indians, said Juett, Hudson's first officer, seemed 'very glad of our comming'.

Hudson is one of the great and tragic figures in the history of exploration, but his recorded life lasted barely four and a quarter years. He was an Englishman. He lived in St Katherine's Parish outside the east wall of the Tower of London. In April 1607 he took communion in the little church of St Ethelburga in Bishopsgate in London prior to an attempt on the north polar route to the East. In June 1611 he was set adrift in an open boat in Hudson's Bay by mutineers. He was not afterwards heard of.

In this brief time he had done more to clarify the geography of the northern seas than any man since the great days of the Vikings. His first voyage, in the *Hopewell*, disproved the theory that a clear water passage ran across the Polar Sea beyond the ice packs. His second, still in the *Hopewell*, dispelled much of the belief in a practicable passage along the northern coast of Russia to the East until the advent of the giant icebreakers. The Dutch East India Company, desperately anxious to find a shorter and an easier route to China, backed his third venture – a proposal to search for a north-west passage. His previous patrons, the Muscovy Company, had by this time decided to switch to the Spitzbergen trade and had withdrawn their support. In 1608 he was invited to Amsterdam for consultation with the cosmographer Plancius – and with his new backers. Almost certainly he took with him a letter and a map from Captain John Smith of Virginia. There were delays and indecision, but in the end the Company decided to provide a ship – and limited funds. They gave him the *vlie boot Halve Maan* – the *Half Moon* – and he sailed from Amsterdam in April 1609.

His recorded instructions are equivocal, but he appears to have considered that he had options: to execute another search for the north-east passage or to begin the adventure of the north-west route.

Hakluyt, the geographer, since the publication of *Divers Voyages touching the Discoverie of America*, had made extravagant claims for a channel through the New World to the East at approximately latitude 40°. Smith's letters and his map dealt certainly with his own explorations but also with the great river – or channel – that ran into the Sea of Verrazano just south of 41°.

How much of this was discussed with Plancius or with the Dutch East India Company is not recorded. In the event, the *Half Moon* stood north to the Barents Sea and met the ice off Novaya Zemlaya. Hudson seems to have had greater difficulty with the pack than on his previous voyages. Swiftly it became apparent that his crew was exhausted, in dread of the ice, and increasingly mutinous. Alternatives were discussed.

The prospect of a search beyond Greenland for a north-west passage aroused, understandably, no enthusiasm; the men had had enough of ice. Hudson chose the third possibility open to him – the warm weather possibility. Late in August he took up the survey of the North American coast where Smith had left off, a little above Cape Hatteras. He glimpsed Chesapeake Bay, entered between the Capes of the Delaware, and finally sighted Navesink on the Jersey coast in the early morning of 2 September. On 4 September he sent 'our boate' inshore to fish, possibly (though the matter is strongly argued) off Coney Island.

His examination of the entrance was limited but methodical. He worked his way slowly up with the tides, entered the river (clearly from the first his objective) and on 12 September anchored probably where the foot of Forty-Second Street is today. On the fourteenth, having anchored the previous night near Spuyten Duyvil, he made the remarkable run of thirty-six miles upriver. On the nineteenth he anchored off the present site of Albany.

The river was shoaling rapidly, and he believed that he had reached the effective head of navigation, but with professional thoroughness he sent boat parties upstream and only when they reported that the river had shoaled to seven feet – and fresh water at that – was he prepared to abandon hope in the possibility of a channel through to the East. By 1 October he was down almost at the mouth of the river again, the river that was afterwards to be called the Hudson.

There was small opportunity for a more accurate survey on the passage down. He had acquired Indian trouble. A succession of bickerings and killings ended in a brief but bloody skirmish in the vicinity of Spuyten Duyvil. With the last shot of the skirmish the *Half Moon* dropped downstream two leagues to a bay '. . . on that side of the river that is called Manna-hata'.

On that side of the river – *not* 'off the island' of Manna-hata. It is completely clear that Hudson regarded this stretch of the river bank as a simple extension of the mainland. The narrow entrance to the Harlem River would have appeared to him as no more than the mouth of a creek as he went past it. Manhattan is not shown as an island on the map that was prepared for the East India Company.

Through the day following the skirmish he lay at anchor in thick weather, but on the next he slipped down through the Upper Bay and worked his way, despite difficulties with soundings, through the Narrows.

> By twelve of the cloke we were clear of all the inlet and steered into the mayne sea.

Manhattan, despite Hudson's omissions, was none the less an island. An inshore island. A fluvial island. Not, as in the more customary usage of that word, an island of riverborne sand and mud like those of the Seine or the estuarine islands of the Rhine and the Amazon, but an island carved, none the less, by the action of a river system through geologic ages.

It is a rock ridge thirteen miles in length; its average width approximately two miles; higher in its northern sectors than in the south; and based firmly on metamorphic rock. Its sedimentary schist provides probably the most stable building platform of any city in the world. It was, when Hudson saw it, largely covered with forest oak, but there were wide grassy spaces.

Hudson was not interested.

Despite what he may have thought when he first sounded his way through the Narrows, the land at the mouth of the Hudson river was positively *not* the gateway to the long-sought-for passage to the Far East.

His homeward route stirs odd speculations. He headed by his own account for Ireland – not Holland. He made his landfall in the English Channel – not the Irish Sea. He anchored at Dartmouth in Devon, and from Devon he sent his papers and his charts on by hand to the Chamber of Seventeen in Amsterdam; but he himself stayed at anchor, to be admonished by the authorities for having made a voyage 'to the detriment of his own country'. He was refused permission to proceed to Amsterdam. Less than six months later, backed by a new, London syndicate, he sailed again in the search for the North-West Passage. In the cold inevitability of epic drama he discovered Hudson's Bay, mutiny, and death in an open boat among the ice floes.

Hudson's successors disregarded Manhattan Island and the remainder of the complex of the river mouth. Hendrick Christaensen and Adrian Block, who set up the first trading post on the river, chose a point above Castle Island across the river from what today is the State capital,

Albany. Their objective was the fur trade. But the United New Netherlands Company, granted the trading rights of the discoveries for four voyages, lasted barely long enough to secure a place in history. In 1618 its charter expired, and three years later the Dutch West India Company was set up, this time with adequate capital behind it, to exploit the area. Even the West India Company was not in the beginning interested in the great harbour that Hudson had surveyed. When Cornelis Jacobsen May sailed with thirty Dutch and Walloon families to establish a permanent settlement in 1624, the majority of the first colonists were allocated to Orange (now re-sited across the river and called Fort Nassau) and the remainder were scattered along the Delaware. Almost as an afterthought a small fort was built to safeguard Dutch interests at the river mouth; even this was not built on Manhattan, it was built instead on Nut – or Nutten – Island in the mouth of the East River. The name was changed long after to Governor's Island.

In the year after the settlement four ships were sent out carrying new families, cattle, horses, hogs, and sheep. The cattle were landed at Nut Island. Because it was small, possibly because the channel between it and Long Island was fordable at low water, they were in due course moved, and Manhattan at last enters the big picture – as a cow pasture.

Even as a cow pasture it failed. Twenty of the beasts died almost at once of eating noxious weeds, but the history of Manhattan had begun.

Willem Verhulst, the director of the new settlement, had arrived with instructions to build a village. He selected its site at the tip of the island and began clearance, but Verhulst was incompetent and, worse, unpopular. When the second instalment of settlers arrived, the fort was still building. Pressure was put on Verhulst, and by the winter thirty houses of raw forest logs were complete together with a stone counting house, a flour mill, and a wind-operated sawmill. In the early winter Verhulst was recalled to Holland and his place taken by Peter Minuit.

Minuit purchased the island from the Manhates tribe. The Dutch were, of course, already in possession. The fort had been built. The name New Amsterdam had been agreed. But the Dutch of the seventeenth century were a litigious people and a cautious one. The West India Company wanted clear title, and a transaction on paper – birch bark, it is claimed, was actually used – was executed. The Manhates marked it with a cross.

The Dutch were willing to pay for it. They did, in fact – sixty guilders – twenty-four dollars – in trade goods. The transaction is still subject to criticism.

New Amsterdam lasted forty years. On 20 August 1664 Colonel Richard Nicolls, writing from H.M.S. *Guerney*, 'riding before Nayack', said bluntly:

> . . . in His Majesty's name I do demand the town, situated on the island commonly known by the name of Manhattan.

His Majesty was Charles II, and being, as the letter further stated, 'tender of the effusion of Christian blood', he offered to confirm the settlers in their life and liberty – and even property. He did take away their name; New Amsterdam became New York.

The Dutch seem to have accepted capture with a stolid equanimity. Nine years later they were recaptured, and accepted that also with a stolid equanimity. Six months later they were finally ceded to England. It must have seemed by this time relatively unimportant. All that they wanted was to be left in peace to trade. Like Paris, Manhattan was already a trading island, drawing by a navigable river on a vast and developing hinterland.

The white settlers of the province at the time of the capture totalled about ten thousand. Like Venice, it was also experimenting tentatively with the possibilities of island empire based on the sea and its own growing capability in ships.

When the revolutionary war began it was not regarded as a model colony. It had, in fact, a long and energetic history of noisy resistance to authority. By the time the Liberty Pole was set up in 'the Fields' there was a mounting catalogue of rebellion. It even threw its symbolic chests of tea into the harbour – four months later than Boston's perhaps, but still significant. Tea was trade, and behind the popular enthusiasm was the fury of the leading merchants against the Navigation Acts.

It does not stand high in the list of defended islands. Neither General Howe nor General Washington performed any notable military miracles in the brief campaign, and the town itself was largely devastated by fire. But it had its reward. On 30 April 1789, George Washington was inaugurated at Federal Hall at the corner of Broad and Wall Streets as the first President of the United States. Unlike Venice at the inauguration of the Doge Foscari, New York frugally limited its celebrations to a single day, but for a year and a quarter it was the capital of the United States. Then, like Paris, it lost honour. Washington moved to Philadelphia as Charlemagne had moved to Aix-la-Chapelle. It does not appear to have offended the inhabitants of New York unduly. Like the original Dutch, they had their eyes most firmly fixed on trade. Like Paris and Venice both, it prospered.

At the end of the first quarter of the nineteenth century it was firmly established in modernity – it had a population of two hundred thousand people and it acquired a traffic problem.

It retains it.

By the middle of the century it had matched Paris in its spread: its *rive gauche* was the great congeries of Brooklyn and Queens, its *rive droite* (more difficult of access because of the nature of the river) Jersey City and the growing towns of the New Jersey hinterland. It had established shipping lines that spread not only to South America and to its own west coast but across the world. It was the entrepôt of North America. Each year, each month of each year, it spread and spread until in time it filled its island – not a little island in mid-river like the Ile de la Cité, but the thirteen-mile length of the rock ridge of Manhattan. At this point it took, consciously, the decision to grow upwards.

Experts quarrel over the precise point in time that marks the birth of the skyscraper. The basic essential of the skyscraper is the steel frame. In a sense the first in the world was that of the Crystal Palace, built by Paxton for the Great Exhibition of 1851 in Hyde Park. It proved the possibilities of an iron structure in which conventional walls played virtually no part in the support of the structural load.

The stimulating mythology that has grown up about the skyscraper suggests that it evolved because Manhattan is an island; congested by building, it had nowhere to go except up. This is, of course, engaging nonsense. The skyscraper did not, in reality, even begin in New York. W. L. B. Jenney designed the Home Insurance Building on the wide levels of Chicago in 1885 for ten storeys, and added two for good measure later. New York followed suit swiftly, not because there was nowhere else to go but because the skyscraper – the gigantic tower on the small plot of land – solved problems of propinquity, equations of economics and audacities of ostentation. There has been over the years a fascinating series of sceptical reassessments. It was said at one period that the addition of a single storey to the existing buildings of the city would have solved all New York's problems. Eight years ago it was solemnly claimed that if all buildings on the island had an average height of seven storeys, there would be no need for 1,000-foot towers. With the upsurge of the fifties and sixties, that average must today be adjusted by a storey or two. The calculations are a matter of logic; they omit, however, corporate emotion.

The skyscraper exists in New York because the panache of New York demands it, because in its advent is quantified the restless energy, the arrogance, the ambition of an audacious city.

Psychologically there is in this nothing new. The Bolognese were busy building upwards early in the 1100s. The Torre Asinelli in Bologna, after all, is 320 feet high and, even if it is four feet out of the perpendicular, the Bolognese have had a deal of use out of it. Venice, if the island argument is to be used, lifts up its own towers to the sky. The campanile of St Mark is 325 feet high, and construction began 1,000 years before the Empire State was dreamed of. There are mud villages in the arid wadis of the Hadramaut that have been skyscraper villages since time began. The skyscraper, still railed at in Europe, is in reality as old at least as Europe and as natural to the development of a city in the twentieth century as a medieval wall or an arcaded piazza to the cities of their day.

Problems remain unresolved. The problem of the self-engendered wind that blows about the bottoms of skyscrapers so that the swirling paper and the cigarette cartons are out of context with the exquisite aloofness of the peaks. The problem of internal human movement that was only partly solved by the development of Mr Otis Tuft's Vertical Screw Railway in the Fifth Avenue Hotel. The problem of external movement, even less capable of solution now apparently, than it seemed in the traffic jams of the 1820s. The problem of squalor and crime and violence in the canyon beds below the peaks. The progressive problem of the adjustment of the human scale to gigantism.

One of the least is the problem of how to see New York.

The skyline, of course, still exists. Hopefully the Staten Island ferry will still reveal its astonishing version of the Downtown section; there will always be, one imagines, that sudden, breathless apparition of pinnacles over the reeds of the New Jersey marshes. But in Central New York itself so much has climbed in recent years that it is no longer possible to see the island for the towers. There are, none the less, two superb viewpoints. The first of these is the bubble of a helicopter 2,000 feet above the Pan-Am building. This is not perhaps wholly perfect but the position at least has the special merit of making it unncessary to see the Pan-Am building itself – barring, of course, the coffin-top of its summit. From it everything radiates. The great line of Park Avenue is the spine of New York and the streets cross it like ribs. Everywhere the towers rise. They are no longer like a palisade as once they were. They are not even like a forest. They are too slender for mountain peaks, too perpendicular. There is almost nothing in nature that can provide the just comparison, nothing except a pale, steel-coloured mineral called, incongruously, stibnite, that is found near Ichinokawa in Japan. It exists as complexes of long, slender crystals, prisms; some of them truncated like the ziggurat skyscrapers that flourished under a temporary and regrettable interpretation of New York's building laws; some slender and pointed as the tower of the Empire State; some square-ended and blunt or sloping and angular. Its crystals interlace, leaning sometimes from the perpendicular, so that it requires a movement of the helicopter and the moment of visual disorientation that accompanies it, to make the comparison perfect. None the less, given the contingent circumstances, New York looks like a mass of stibnite crystals.

There is only one better viewpoint, which also requires certain concatenations of air circumstance. A long while ago and a little after sunrise I flew south-west at 20,000 feet across Long Island. It might have been somewhere above Jamaica Bay, we could not tell. The land and the sea alike were covered by a mist, a shallow mist, not above 300 feet in depth, and coloured like a pallid cyclamen in the morning light. Out of that colour rose two fantastic groups of towers, the main massifs of Manhattan Island. They were in that early sunlight utterly clear, infinitely precise, superbly aspiring. They had no contact with reality, no element of the substantial. They were the tops of the towers of Ilium.

Below them, though we could not see it, there was also Brownsville, the newest symbol of urban perdition. The mist that hid it was an intricate and exquisitely evolved commixture of carbon monoxide, lead compounds, fly ash, sulphur dioxide, hydrocarbons, nitrogen oxide, asbestos particulates and trace elements of a wide variety of other noxious substances. A similar mist observed over Los Angeles by the Apollo 10 astronauts from 25,000 miles out in space was described as a cancerous smudge.

War

MALTA

> In the night the gods washed me up on the Isle of Ogygia, the home of the fair Calypso, that formidable goddess with the woman's voice. . . .

Ogygia may not be Gozo, lesser of the Maltese islands; Ramla is possibly not the beach where the shipwrecked Odysseus drifted at last ashore; the cave which the people of Xaghra show you proudly has probably no association with any nymph. Yet for 2,000 years Gozo and Malta were the Islands of the Mother Goddess, and it is upon the race memories of those years that Homer drew for the seven-year enchantment of his heroic lecher.

As has already been said, the track of the *Odyssey* is the despair of navigators. Odysseus, according to his account, was driven by a westerly gale from the Island of the Sun – tentatively identified as Sardinia – following the slaughter of the sacred cattle. His ship dismasted and shattered by lightning, Odysseus, unaided, lashed the great timbers together and rode the wreckage back through the strait of Scylla and Charybdis between, by firm tradition, Sicily and Italy. Logically he then passed into the Ionian Sea. Malta and Gozo lie on the edge of that sea. Despite the danger of applying logic to the *Odyssey*, he could have ended that most perilous adventure in Calypso's bed.

Possibly there never was a Calypso – there may never have been an Odysseus – there *was* a mother goddess. She exists still in half a hundred statues on Gozo and on Malta – formidable, peacefully malevolent, obese to the limits of obscenity.

Time-tables in pre-history are at least as uncertain as the track chart of the *Odyssey*. Even with Carbon-14 tests no one today is bold enough to defend an absolute schedule, but human remains exist in the upper levels of the great Ghar Dalam cave above the levels of elephant bones, that have been dated at 3800 B.C. Skeletons of an offshoot of the slow wave of early Neolithic people who flooded across from Calabria to Sicily, they had found a green island, forested and fertile. They must have been a boat people – they brought cattle with them – they lived in caves, they had no weaving, their pottery was primitive, and they were weaponless. Though metals came, life persisted in an engaging simplicity. Nothing was made in the first 1,000 years that can be considered warlike.

At the end of that time began, with the building of Ggantija above Calypso's bay, the tremendous efflorescence of the great temples. It is the largest of thirty major buildings – megalithic, constructed of huge blocks of masoned stone, some of them of as much as fifty tons. Initially trefoil-shaped structures opening from a roofless central court, these temples developed by way of the enormous underground galleries of Hal Saflieni to the magnificent quadruple group of Hal Tarxien.

Little is known of their builders, almost nothing of their religion save that with its brooding statues it was an expression of the female principle. One thing only is certain – they vanished. There is no evidence of cause, nothing to suggest a massacre, an epidemic, even a prolonged drought. They ceased to be, but left behind them an enduring legend of an island and a goddess—and Homer heard it.

A Bronze Age people replaced them, a people with weapons, and at Borg in-Nadur on Malta they built, across the base of a promontory, a simple wall that is historically the starting point of the most complex island defensive system in the world.

If the pre-history of Malta is complex, its history is frenetic. Lying in the gut of the channel between Europe and Africa, centred between the basins of the Mediterranean, it was the strategic nexus of the classic sea world.

The Phoenicians grasped this first and held Malta for six centuries to cover their trading ventures. At the end of that time the strategic polarization changed to cover the north–south grain route between Africa and Rome. Intermittently it shuttled between the sea powers until, in natural process, it became the base for the ultimate expedition against Carthage. For the next 1,000 years it was nominally under Rome. As Rome diminished, the Goths raided it; the Vandals followed them and history becomes a disjointed chronicle of conquest and evacuation.

Easter Island called itself, in overweening pride, the Navel of the World; Malta *was* the navel of the Mediterranean. Less than sixty miles from Sicily, 190 miles from Africa, 1,000 from either the Straits of Gibraltar or the Levant, it offered to whoever seized it a stringent control of the inland sea.

The Arabs took it in the process of making themselves a sea people. In the course of their conquest of Sicily a forgotten Byzantine garrison fought Malta's first great siege, survived a desperate year, and was lost. Two hundred years later the Normans took it from the Arabs. They had come south not as Vikings in longships but as mercenaries of the Lombard cities, and it was they who drove the Saracens from Sicily, and it was from Sicily that Roger the Norman captured Malta and began again the process of brief sovereignties as the control of the narrows passed from hand to hand.

It was a barren island now. Neolithic man had begun the destruction of its limited forest. Bronze Age man completed it. Siccation is a cumulative evil – the gregale blew out the dust of the dried earth, the rains washed the soil itself out to sea. In a sense the siege of drought was its longest siege of all, and the existing field system of terraces and walled plots is, in its intricacy, the token of victory. But when in 1523 the Commission of the Knights of the Order of the Hospital of St John of Jerusalem examined it, its members permitted themselves no illusions.

> 'The island is a rock of soft sandstone . . . barely covered with earth which is likewise stony, and very unfit to grow corn . . . except for a few springs there is no running water . . .'

The harbours, however, were magnificent.

The Order of St John began as a hospital of the Benedictines in Jerusalem. For its assistance to the First Crusade it was expanded in the general interest of pilgrims. Its vows were to chastity, obedience and poverty. For the protection of pilgrims it became a military order, and in the bizarre history of the Holy Land it clashed with that other purely military order, the Knights Templar. Both became rich and politically powerful, both were swept out of the Holy Land in the last disasters of Outrêmer.

The Order of St John seized Rhodes from Christian Byzantium, made itself a sea power, and became the bulwark of Christendom against the Turk. It achieved in the process incredible wealth. By 1500 it was believed implicitly that The Religion, as it arrogantly called itself, was as rich as the Church of Rome. Its vow of poverty was forgotten, its obedience conditional except in war, its chastity cynically interpreted – fortitude remained.

In 1522 Suleiman the Magnificent decided that it must be destroyed. For six months the Order held Rhodes against a fleet of 400 ships and 200,000 men. On Christmas Eve Villiers de l'Isle Adam was granted the honours of war and the right of unimpeded departure.

As Grand Master de l'Isle Adam took his Order and his problem to Rome. It was not in any aspect easy of solution. The nations were desperate for a defence against the Turk, but they were almost equally terrified of such turbulent and unpredictable defenders. Past hope of a better choice, the Knights accepted Malta in 1530. There were at least the harbours.

The defence of pilgrim caravans in the desert had been transmuted now into sea raids against the infidel – and such others as the Order decided. It called these raids, in which novices served an often semi-piratical apprenticeship, 'caravans', and it was inevitable that they emplaced themselves instantly at Birgu, the village of Grand Harbour with its dilapidated castle, rejecting Mdina, the ancient capital in the interior of the island. Across the neck of the peninsula on which Birgu stood de l'Isle Adam began the great wall that is called Castille, matching the prehistoric wall at Borg in-Nadur.

In the beginning the Knights built tentatively, cherishing a deep nostalgia for Rhodes, but the Castle of St Angelo and the adjacent peninsula of Senglea were boldly improved. A new castle, St Michael, was built at its point and below the heights of Sceberras on the central finger of the harbour they erected a fortress – St Elmo.

Suleiman left them too long. When, however, in 1565 the clamour of the harem at the capture of a trading ship owned by the chief eunuch drove him to action, he thundered:

> I say now that, for their continual raids and insults, they shall be finally crushed. . . .

In this year Jean Parisot de la Valette was Grand Master and his spies had brought him the word that a vast Turkish fleet was assembling. Everything that he could do to prepare the island for attack was already done; all he could do now was to wait.

Nathaniel Dance's sober portrait of James Cook, the British naval officer who was made a god on the island of Hawaii.

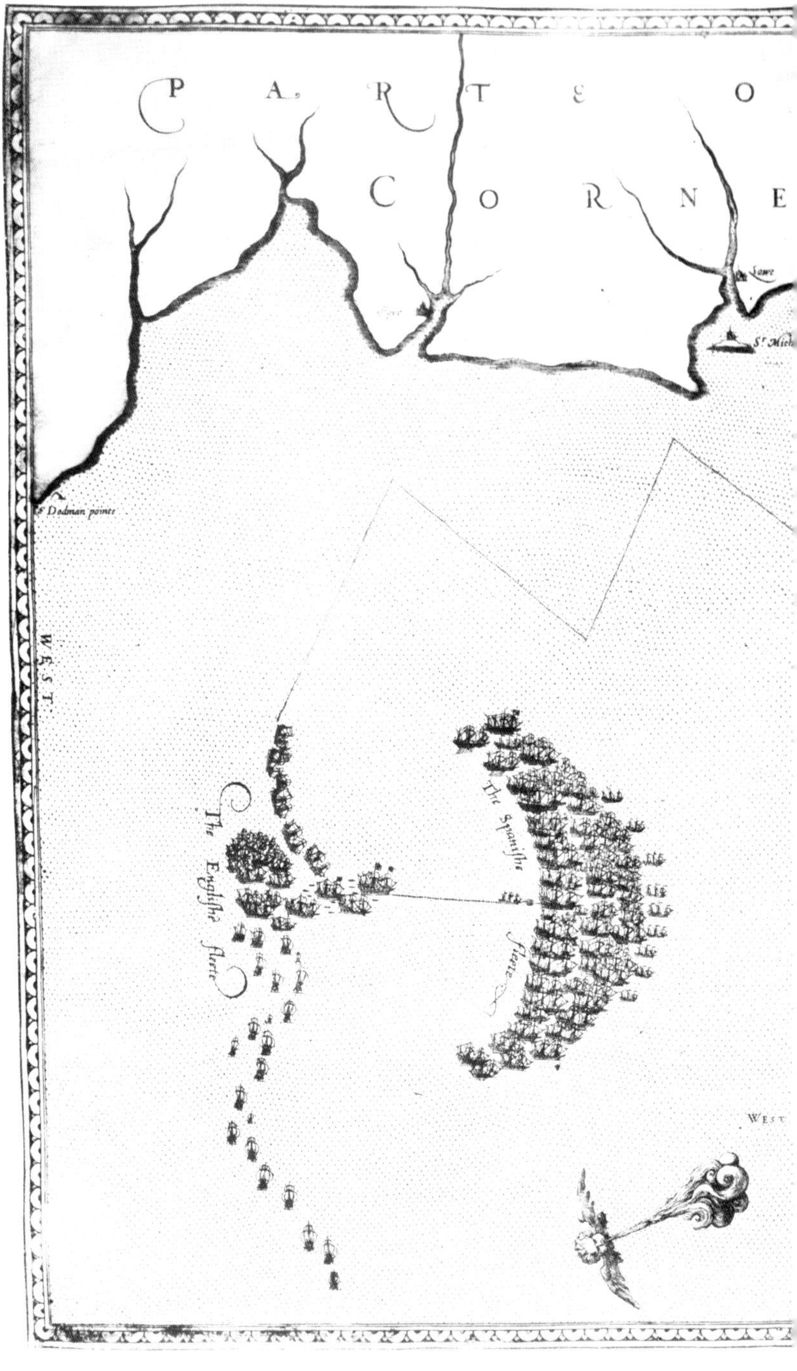

The track chart of 'Howard's challenge', the manoeuvre that gave Howard the weather gauge of the Spanish fleet at the opening of the Armada campaign and made victory possible.

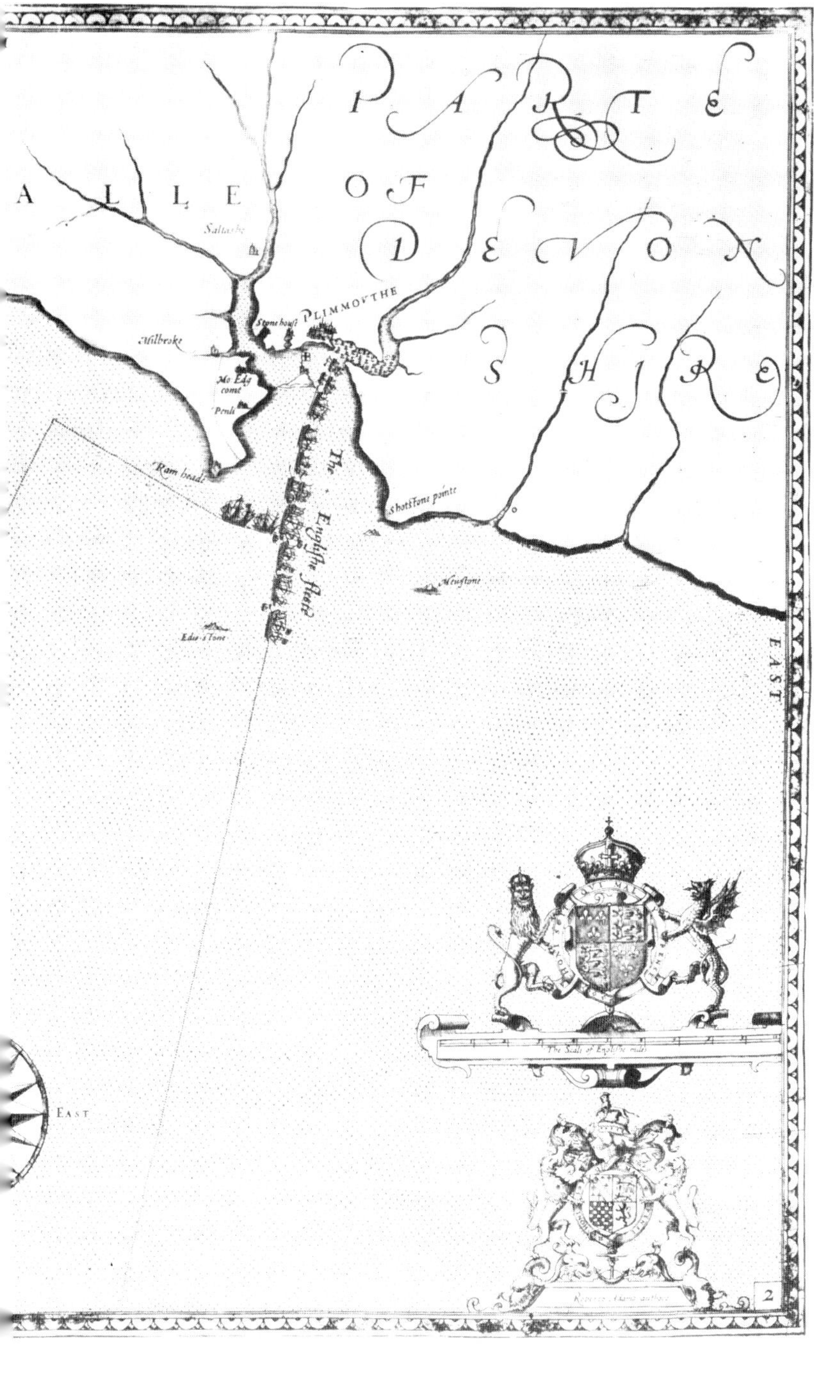
PARTE OF DEVON SHIRE
ALLE
Saltashe
Stonehouse
PLIMMOVTHE
Milbroke
Mo Edgcome
Penli
Ram heade
The Englishe fleete
Shotstone pointe
Mewstone
Edistone
EAST
EAST
QVI MAL
The Scale of Englishe miles
2

Perez d'Aleccio's remarkable study, made for the Palace of the Grand Master of the Knights of St John, of the Turkish attack on St Elmo which marked the opening phase of the siege of Malta.

Suleiman had been told by Dragut Rais, the Greek mercenary who had made himself the greatest captain of the Mediterranean, that Malta's defences were inadequate. Dragut had sacked Gozo eleven years before, five years later he had pillaged all Malta; at Tripoli the Knights themselves had yielded the fortress to Sinah Pasha intact. He persuaded Suleiman easily that his time was more than ripe.

De la Valette himself awaited the attack with a steadfast calm. The Order, aware that the Great Powers had not sent the reinforcements promised, awaited it with something less than its customary faith in God. Early in the morning of 18 May 1565 three shots were fired from the ramparts of St Angelo and the watchers on the heights saw on the misty sea a vast semi-circle of 180 Turkish ships.

It is on the courage and the immaculate psychological perceptions of de la Valette that the issue of the siege of Malta hung. He had already recalled the Knights of the commanderies in Europe. The detached fortresses of Gozo and Mdina were garrisoned. The cavalry was positioned at the latter. The walls at Birgu and Senglea were complete. The bridge of boats that connected them was in place. The great chain across the mouth of Dockyard Creek was ready.

That chain is the key to an understanding of de la Valette's policy. He had long accepted that he would have to fight not a sea battle but an inland siege, for on 18 May only five galleys lay in harbour, so grievously had the sea power of the Knights diminished. These five made a ritual gesture of reconnaissance and returned to harbour. The great chain was winched into place.

The key to an understanding of Suleiman's policy lies in the command structure of his expedition. The ships themselves were commanded by Admiral Piali, his granddaughter's husband; the soldiers by Mustapha Pasha, his leading general. They sailed under injunction to delay their main attack until Dragut Rais joined them. Suleiman was not prepared to permit a concentration of power in the hands of any one man sufficient to be turned against himself.

The effects of this division of authority are clearly apparent in the opening moves. In a singularly meaningless feint the fleet swept past Malta to the eastward, ran up the south coast and anchored off Mgarr. In the small hours it dropped back to the open harbour of Marsaxlokk and landed 3,000 men. Almost at once, in a brief clash with a cavalry patrol, a French Knight and a Portuguese novice were captured. Interrogated under torture, they declared the great wall of Castille to be the weak point of the defence. Ignoring the injunction to wait for Dragut Rais, Mustapha launched a major assault. Six desperate hours later he withdrew from Castille with heavy loss, and the prisoners, for their valiant lying, died under the bastinado.

The siege divides into three principal phases: the opening assault, the siege of the new castle of St Elmo, and the long and terrible investment of Birgu and Senglea.

Withdrawing furiously from Castille, Mustapha proposed to seize the northern area to prevent the landing of reinforcements from Sicily. Piali proposed instead the reduction of St Elmo to permit the entry of his fleet into Marsamxett Harbour. Mustapha lost.

St Elmo had been hurriedly built; it was small, dominated by Mount Sceberras, and the Turkish engineers estimated that a five-day bombardment would carry it. As the guns opened, de la Valette received the first of an endless series of dispatches informing him that reinforcement was again postponed.

The defiance of St Elmo is one of the superb chapters in the history of The Religion. With its bastions crumbling, its dead replaced grimly by night across the harbour from St Angelo, it fought on. The five days passed and still it held. Dragut arrived, harshly criticizing the conduct of the campaign, and built fresh batteries on the headlands of the harbour. Still it endured. On a single day 7,000 shots were fired against it, and finally when Dragut determined it had been effectively softened, the Janissaries stormed it.

In the fury of the attack Dragut was desperately wounded. Already the five days had stretched to ten, the ten to twenty. For thirty-one days St Elmo held out in one of the great defiances of history. Then, in the dawn of 23 June, the Knights who still lived took the sacrament, burned their religious pictures, concealed the sacred vessels, and went out to die. De Guaras and de Miranda, the commanders of the garrison, long since wounded, ordered themselves to be placed in chairs in the centre of the great breach, their swords in their hands. As the Janissaries, the Iayalars and the Spahis swept down from Sceberras, they died there; and as they died, Dragut – the Drawn Sword of Islam – died also of his wounds.

The siege of Malta entered its third phase when the prow of a Turkish galley thrust over the skyline of the ridge between Mount Sceberras and the mainland. In one week eighty ships and boats were dragged by the galley slaves from the new-won harbour of Marsamxett by way of

Pieta Creek across the high ground to the Marsa. It was the dead Dragut's bequest to the campaign. Long before, he had dragged his galleys on rollers across Djerba to confound the famous Genoese admiral, Doria.

The plan was apparent now: an assault on the twin peninsulas by water from two directions, with a simultaneous assault by land on the great curtain walls. Early in July seventy guns, ringing the area from Gallows Point round the harbour to the shattered walls of St Elmo, opened a bombardment that lasted two terrible months.

Under its thunder the attack was launched. For this too de la Valette had prepared – there were water obstacles as well as walls, stakes driven into the harbour bottom, chains, booms, ships' masts. The main attack began with attempts by swimmers from Corradino to wreck the water obstacles. Maltese swimmers repulsed them. The defences were intact when the ships drove in with heavy loss, but a chance shell fired a powder magazine on Senglea. The sea wall was breached and through the gap the Turks surged in. Simultaneously the assault opened on the landward side.

Judging his opportunity with a cold brilliance, de la Valette counter-attacked across the floating bridge, and as he did so ten large craft crowded with Janissaries were flung into the assault against the head-land of St Michael. Nine were sunk, 800 Turks were drowned, the attack from the land was beaten back, and Senglea was secure again.

All through the month the guns thundered and no help came. One of the first strengths of an island is the water barrier of the sea. One of its first weaknesses is the difficulty of reinforcement across the sea. The isolation of an island is a special peril, political as much as physical. By early August the Turks had gathered their strength for another and a greater storming, and still there was no help from Sicily or Spain.

At Birgu the outer wall was breached, but an inner wall held and the slaughter was enormous. Senglea was breached again, the outer walls of St Michael were carried, and de la Valette, looking across the mêlée from his own harsh battle, could see the Turkish pennants on the citadel. It was the crucial moment of the siege. There were no reserves now – no one to send across the floating bridge.

And in this instant of time the incredible happened: above the thunder of the battle de la Valette heard the Turkish trumpets sounding the retreat.

Behind the low hills past the Marsa the cavalry from Mdina had carried out a devastating attack upon the Turkish camp. The wounded, the sick, were slaughtered where they lay. The camp was burned; stores, spare powder, baggage, everything was burned. In the height of the confusion Mustapha was told that a relief force had slashed its way across his rear.

There was no relief force; only the cavalry, and a messenger from Garcia de Toledo, Governor of Sicily, postponing relief once more.

Mustapha turned to mining, the bombardment re-opened. On 18 August he blew a vast charge under the wall of Castille, and through the dust of the explosion the Turks attacked. Seizing a pike, armourless, and with the bells of the cathedral ringing over his head, de la Valette led his staff into the new breach – 'How is it possible for a man of my age to die more gloriously . . .?'

He did not die. The Turk was driven back again. But as new siege engines were brought to bear and the shelling increased, de la Valette was urged to retreat to the citadel of St Angelo. He refused gravely. The Knights and the people of Birgu, he said, would meet their end together. Even de la Valette believed that the end was near.

It was his one misjudgement. Dysentery and enteric had decimated the Turkish army; death in battle brought it to danger point. And on 7 September something that was almost a miracle happened: Don Garcia de Toledo, with a reinforcement, landed. On 8 September the bells of the Cathedral of St John's rang once again – for joy.

Defeat at Malta halted the westward thrust of the Turk. Defeat at Lepanto six years later broke his sea power.

In its origins and in its course the siege was determined by the characteristics of Malta as an island – its vulnerability to surprise attack, its problems in relation to reinforcement, the absence of the option to retreat. Under the leadership of de la Valette, The Religion triumphed against circumstance.

By its triumph it was again a power. But when, six years later, Lepanto crushed the Turk at sea, it failed to recognize that the reason for its being – and therefore the reality of its power – had disappeared. On Malta, secure in glory, rich with the recognition of Europe, arrogant, aristocratic, authoritarian, it embarked on a madness of fortification. Members of the council who wished to abandon the island were overborne. The repair of St Angelo and the landward defences was begun at once. Francesco Laparelli, Michelangelo's assistant, was sent by the

Pope to plan a new Malta, and on the perilous heights of Scebarras the city of Valetta was built. Matteo Perez d'Aleccio, who had worked with Michelangelo on the frescoes for the Sistine Chapel, painted the extraordinary semi-primitive semi-Renaissance designs now in the National Maritime Museum at Greenwich for the decoration of the Palace of its Grand Master.

The island of Manoel was fortified in turn; on Dragut Point the fortress of Tigne was built; on the eastern headland rose Fort Ricasoli; Cospicua was walled; inshore of Valetta the great secondary system of Floriana was designed with outworks covering the track of the galleys from Pieta Creek. A hundred years later they were still building the fantastic extravaganza of the Cottonera lines. They were at the point of completion when, on 10 May 1798, The Religion surrendered to the young Napoleon without firing a shot.

The blockade of the garrison that he left to hold it ended with Britain in control of the strategic fulcrum of Mediterranean sea power for a century and a half.

On 10 June 1940 Mussolini declared war against the Allied powers. On 11 June ten Italian bombers attacked Grand Harbour. On the following day the Italian submarine *Bagnolini* sank the British light cruiser H.M.S. *Calypso*. Somewhere in Malta someone that day must surely have murmured: '*Absit omen*'.

In the view of the British Army Council and the Air Staff it was 'impossible to defend Malta effectively'. The Official History says that the Naval Staff accepted reluctantly that 'in the event of war with Italy the Mediterranean Fleet would not be able to use Malta as a main base'.

Much of military policy in 1939 was based on misjudgements of air potential made fashionable in pursuit of the Trenchard doctrine of air power between the wars. Malta, though the Sicilian airfields were barely twenty minutes' flying time away, was in fact defensible, could have been adequately prepared against the Italians, and did survive even when the harsh weight of the Luftwaffe became a vital factor in North Africa. Its part in the invasion of Europe through Sicily and the ultimate collapse of Italy was won by enduring a siege longer and incomparably more bitter than the great siege of the Knights.

The word itself has a startling variety of meanings. In relation to war it means 'investment to cut off all communications'. Malta was invested from the air with the first attack by the Italians on Grand

Harbour. The great ramparts of the Knights were obsolete, the Victorian defences of the naval base were useless. The island had been permitted no air defence at all.

It improvised one.

When the Mediterranean Fleet withdrew from Grand Harbour to the distant and temporary security of Alexandria, four Sea Gladiator aircraft from H.M.S. *Glorious* were left encased at the seaplane base of Kalafrana. They were acquired something less than officially by R.A.F. pilots, mostly from heavy flying boats, and after sketchy training, made ready for combat. Despite official hesitations, they took off on 11 June as the first bombers of the Regia Areonautica swept over the island. Malta christened them, with a rare felicity, Faith, Hope and Charity.

Historic parallels are often easy and always dangerous, yet certain distinct similarities exist between the siege of the Turk and the siege of the air.

Like Suleiman, Mussolini attacked because, after 400 years, Malta was again a threat athwart vital lines of communication – once more those of the north/south line between Italy and Africa as in the days of Rome and Carthage.

Like the Knights' siege, the assault which began in 1940 divides into three main periods. The first of these covers the three months of the initial Italian attack. Ill judged, as crassly optimistic as Mustapha's frontal assault against Castille, it failed. The second began with the arrival of the Luftwaffe at the Sicilian air bases in January 1941. Like the Turkish campaign after the arrival of Dragut Rais, it opened with a crescendo of expert fury. The aircraft carrier H.M.S. *Illustrious*, covering a convoy to the island, was badly damaged by dive bombing. In desperate need of repairs, she berthed in Frenchman's Creek, and above her the air battle developed in new and fantastic levels of violence. She survived. Her damage was made good by sheer guts, and she slipped out triumphantly; but over Malta the baulked Luftwaffe poured its wrath. By June the state of the island was perilous. The Wellingtons of the slender bombing force had been withdrawn, the reconnaissance Sunderlands had gone. Despite heroic attempts at reinforcement, the fighters were lamentably weak. The submarines that had lain submerged in the harbour by day had been pulled out, and the small destroyer force had sailed to take a harsh part in the disaster of Crete. On the African mainland General Wavell had begun the retreat from

Cyrenaica. To those who endured it seemed that the Luftwaffe, like the Turks at the crisis of the storming of Senglea, was at the edge of triumph. And on that edge, like the Turks at Senglea, it was withdrawn abruptly – to the Balkans for the opening of the campaign against Russia.

The third period began when, as the Eighth Army opened its winter offensive against Cyrenaica in December 1941, the German command, despairing finally of effective Italian assistance, moved back the Luftwaffe to cover Rommel's lines of communication. As in the last phase of the Turkish siege, a ruthless methodical application of power came most desperately close to triumph. In April 1942, 400 sorties were flown against Malta in a single day, 6,728 tons of bombs were dropped, east and west of the island the convoys were beaten back. We who sailed with them could measure the terrible necessities of that time in the magnitude of our own disasters. At Berchtesgaden Hitler and Mussolini determined to invade early in July. Operation 'Hercules' was planned to capture Malta. Malta fought on.

And then between April and July in a welter of Italo-German mistrust stimulated by the magnificence of the defence, plans for invasion foundered. No attack was launched.

In November 1942, with Montgomery halfway across Tripolitania and the First Army deep into Tunisia, the siege was lifted. From Alexandria a great convoy steamed unopposed into the Grand Harbour and on the bastions of St Elmo, in the bomb-shattered Baracca Gardens, on the walls of the Three Cities the people of Malta massed to cheer it in.

The third of the great sieges was over. The price paid. 14,000 tons of bombs had been dropped on the islands, 24,000 buildings had been destroyed or damaged, 1,468 civilians killed, 568 aircraft lost in their defence; two aircraft carriers, five cruisers, a fleet of destroyers and submarines and small craft had been sunk fighting the convoys through.

Malta's loss was heavy. The loss of the Axis powers was absolute. In the course of the air siege they had sacrificed 1,129 aircraft – in the loss of the battle they had lost Africa.

Malta is restored. The scars of war today are hardly anywhere apparent. The castle of St Angelo is whole again. The rock-cut church of St Mary in the fortress, it is true, is altogether wrecked. In Birgu three of the earliest of the auberges of the Knights – Italy, Germany and Leon – are gone, but the streets have long since been cleared of the choking rubble and the houses are homes again. Valetta is rebuilt, though the auberges of France and Auvergne were destroyed. Mdina shows no wounds. The honey stone of the rebuilding has merged into the antique landscape. Yet Malta is wholly changed.

In an important sense the history of its long association with the Royal Navy ended with Admiral Cunningham's signal to the Admiralty on 11 September 1943:

> Be pleased to inform Their Lordships that the Italian battle fleet now lies under the guns of the fortress of Malta.

The change in the island's relationship with Britain is less easy to place in time. At some unremembered moment in the siege the small light of independence, scarcely more than a glimmering through the 2,000 years in which the island was a prize between the warring sea powers, became a flame. It is perhaps that moment which is acknowledged in the simple words signalled to Malta in the crisis of the battle:

> To honour her brave people I award the George Cross to the island fortress of Malta to bear witness to a heroism and devotion that will long be famous in history.
>
> GEORGE R.I.

It is these same brave people who achieved their independence. It is their task to preserve it. Malta still lies between East and West, between North and South. Like the Cretans and the Greeks, the Phoenicians and the Romans and the Turks, Russia is extending her strength in the Mediterranean. Malta is still a small island between contending powers. After 2,000 years she has only her virtue as an island – and a proven capacity for heroic endurance. She exists delicately in the imbalances of the twentieth century world.

The election of June 1971 and the victory of the left, led by Mr Mintoff, has for the West much of the significance of the three shots that were fired on the sighting of the Turkish fleet in 1565.

BRITAIN

The Flandrian Transgression embodies in its name all the sonorous overtones of a papal denunciation of mortal sin. None the less it is no more than a label for a moment in topographical time – the long moment when the rising of the waters of the Atlantic which followed logically the melting of the northern ice-cap flooded the hollows of the North Sea basin. The Dogger Bank was swallowed about 7000 B.C. Two thousand years later the lowlands that lay on the coasts of Germany and Holland and Belgium were engulfed. A little after, in a final surge, it may be in the vigour of a summer storm, a wave broke over the diminishing chalk ridge between the peninsula of Britain and the Continent, and Britain became the archetype of the offshore island.

It did not require to be discovered. Long before the sea established its final level, sparse waves of Stone Age people had followed the rising temperatures to find a living in the raw peninsula. Long before the year 2000 B.C., when the Transgression came to its end, Neolithic man had moved in.

It is improbable that even in the earliest stages of the scouring of the Straits of Dover there was ever any real severance of communication between the new-made island and the mainland. Certainly with the arrival of Neolithic man there was no problem. He was a hunter and a fisherman as well as a sower of crops. His middens hold traces of cod and plaice and deep-water fish. He must have had effective boats. The sea history of Britain begins with him.

In the beginning it is, understandably, discontinuous and obscure, but very early there are indications of traffic not only across the narrows of the Straits of Dover but across the wide mouth of the channel from Brittany to the West Country. By tradition there were, with the beginnings of the tin trade, longer voyages – the ultimate stages, possibly, of the great trading ventures of the Phoenicians, which had already brought to Malta the opening of history and had made Cadiz the entrepôt of the Atlantic coast. Certainly by the time the Romans came, half a century before the birth of Christ, there was an established sea trade.

The island was settled and beginning to be populous. Successive migrations – Bronze Age man followed by a series of tribes of the Belgic peoples – had flooded the south of the island and swept far into the northern hills. Already there was a network of tracks that were embryo roads. There were trading ports and fisheries, villages and towns, and the great hilltop earthworks of Iron Age man. But there were, as far as acceptable evidence exists, no ships of war.

The sea war history of Britain begins with Julius Caesar. In an amphibious operation of an astonishing incompetence, Caesar landed an army of two legions on the beach of what is now Walmer or Deal in 55 B.C. His cavalry had failed to button up with the fleet off Ambleteuse. His fighting ships were carelessly drawn ashore on the unsheltered shingle. His transports were anchored in shoal water. The top of a spring tide and an onshore breeze – it cannot have been a gale since his horse boats weathered it without loss – coincided to wreck his ships and his plans.

Caesar withdrew, claiming that he had intended no more than a reconnaissance in force. The following year he compounded his ineptitude. Despite the existence of a safe harbour in the channel of the Wantsume, barely ten Roman miles to the north, he used the treacherous beach again. Two nights later for the second time the fleet was driven ashore, desperately battered. Speculation as to the degree in which a small but competent sea force might have interfered with the invasion in the face of this displayed inadequacy is profitless; there was no British sea force.

There was still no organized sea force when Aulus Plautius landed a century later, nor were there through the years ships enough to challenge the Roman fleet that supported Agricola's right flank in the campaign to Mons Graupius and that sailed round Scotland in the final stages of the conquest.

The character of Britain as an island altered drastically with the Roman conquest. From being an independent offshore island it became an appanage of a continental empire.

The Romans were not a maritime people. They had developed the possibilities of the sea within the limits of the Mediterranean situation largely by the employment of subject peoples, but they had not acquired – they had not prior to this needed to acquire – a concept of oceanic sea power. They perfected, for the uses of the new province, the trading potential that had existed before they came and they produced for it an escort system. When in due course the Saxon threat developed on the North Sea coasts, that escort system was elevated into a fleet. It was not based on Britain but at Boulogne in the narrows of the Channel; its task was not only to safeguard communications with Britain but to protect the Channel coast of Gaul.

It is one of the major ironies of the history of Roman Britain that it was this fleet which first severed effectively the connection with Rome. Carausius, a Menapian from the Low Countries, commanded the Channel area. Accused of hi-jacking the Channel pirates he was supposed to suppress, he revolted against Rome, seized the coast of Gaul as far as the Seine, and took over Britain. Maximian's attempt at reconquest with an improvised fleet was – almost inevitably – broken by bad weather, and Carausius was recognized as one of a triumvirate of emperors.

Edged out of Gaul, he set up a virtually unassailable position in Britain, and the great chain of the fortresses of the Saxon Shore which ends magnificently at Porchester may, according to modern American theory, have been built not against the Saxons but to keep out the revenge of Rome, with the Channel as its moat.

How Britain would have developed had Carausius lived is tantalizing to consider, but he was assassinated by his Finance Minister, Allectus, and Allectus, altogether without feeling for the sea and the realities of sea potential, was destroyed.

Carausius's brief reign remains the first expression of genuine sea power in Britain. Long before Alfred, its traditional father, he was the founder of a British Navy.

Rome in due time decayed, Western Christianity expanded – largely among the offshore islets of this offshore island – and across the peace of the Holy Isle of Lindisfarne in 793 broke the opening tumult of the astonishing maritime explosion of the Northmen. Spreading with the years to Vinland in the west, reaching to Byzantium in the east, absorbing Malta in the south, it shattered the simple pattern of the coasts of the known world.

It originated in raiding and developed into settlement. The Saxons, the Anglii and the Jutes, who had followed the withdrawal of the Romans and settled England, moved almost imperceptibly from intermittent defence against local attacks into a continuing war against Danish incursion.

It is a conspicuous element in the history of England that its conquering peoples, reaching the island by sea and obtaining lodgement on it by virtue of sea power, have tended to lose in the course of settlement their capacity to utilize its possibilities. Not until Aethelwulf's 'King's Ships' gave battle with fleeting success to the Danish invasion was there any attempt to organize an effective defence upon the sea. Alfred – Aethelwulf's son – himself turned to the sea only after he had shattered the armies of the invaders in his land campaigns and then, primarily, in hope of breaking up fresh concentrations of the enemy:

> He commanded to be built against the Danish warships long-ships which were well-nigh twice as long as the others.

On the *Anglo-Saxon Chronicle*'s statement rests Alfred's claim as founder of the British Navy. The ships themselves were unsatisfactory. Moreover continuity was not established. It was a century before Aethelred the Redeless assembled another fleet against a new Danish threat and, says the *Chronicle*:

> We had not the good fortune nor the worthiness that the ship force could be of any use to this land.

It was of little 'use to this land' fifty years later against William of Normandy.

It should have been. Harold's handling of the double threat of Harald of Norway in the north and William the Bastard in the south is marred by a singular failure to grasp either the advantages or the inherent dangers of an island. Through the summer of 1066 he stood along the south coast of England with his housecarls and the fyrd, the conscripted militia. His King's Ships and his hired vessels operated between Sandwich and the Isle of Wight. The northern threat he left to the great earls, Morcar and Edwin.

Across the Channel William built ships and gathered men. Clearly his intelligence was accurate and swift. He waited for the Norwegians to make up their minds before he moved, he timed his own sailing in the knowledge of Harold's difficulties.

Delay had exhausted Harold's supplies. The men of the fyrd, desperate to get in the harvest, were restive. The ship men were unruly. He altered his plan abruptly, gambling on the probability that the Norwegians would attack first, moved the housecarls to a strategic position at London, disbanded the fyrd, and ordered the fleet to the Thames.

The movement of the housecarls was altogether justified in the light of the dual threat; the disbandment of the fyrd, restless through it was, was an error; the re-disposition of the fleet was an unmitigated disaster. It was the decision that cost Harold his throne and his life.

The timing is not everywhere clear, but four days after the withdrawal of the English fleet William moved his ships and his base from Dives near Caen to St Valéry-en-Caux in striking distance of the Sussex coast, losing some of his ships in a gale on passage. In the first half of September Harald raided the Durham and Yorkshire coasts unopposed, worked up the Ouse to Riccall, and at Fulford crushed Edwin and Morcar.

Harold, racing north with the housecarls, made contact at Stamford Bridge on 25 September, destroyed the victors and, almost without pause, turned south again to forestall William. But William, seven nights after Fulford, with Harold safely 250 miles to the north, had made his move. Sailing as guide of the diminished fleet, a lantern at his masthead, he outstripped his main body. There was panic in his own ship, disorganization in the fleet. Resolute interception anywhere on passage would palpably have disrupted his plan, harassment at the landing could have delayed the build-up of strength at Pevensey.

There was no interception. In all the Channel there was no English fleet; it is one of the moulding errors of sea history. Harold – the evidence is manifest – saw his problems in terms of the land campaign. The fleet lay useless in the Thames.

The consequences of the Norman invasion are a precise antithesis of the revolt of Carausius. Where the Roman admiral established an insular independence, the Norman Duke re-absorbed Britain into the Continent as the appanage of a dukedom, barely disconnected, almost visible from the Norman coast at Le Tréport. Control of the Channel reverted to the Roman practice before Carausius, short-sea shipping maintaining escort and communications.

William himself regularized the system by granting trading privileges early to the three ports of Sandwich, Dover and Romney in return for the provision of ships to cover the Channel traffic and to serve with the King's Ships in war. In due course the Cinque Ports complex covered the south-east coast from Pevensey to Brightlingsea in Essex.

The Cinque Ports adjudge themselves collectively the fathers of the British Navy. As has been suggested, that honour belongs to earlier acts of procreation; none the less they served an important purpose in an inchoate time.

They provided often more than the fifty-seven ships agreed under their charter, and their fleets, under their own admiral, had a discipline and a cohesion that were not apparent always in the ships chartered from individual merchants. Their primary weakness lay in an almost outrageous independence that was legalized under Henry III when they were given licence to harry the French coast at will on the basis of a profit sharing arrangement which gave one-fifth of the proceeds to the King. They maintained their own private wars, they quarrelled as ferociously with Yarmouth and with the West Country men as they did with the French. In just return, counter-raids laid their ports waste, their ships were burned or captured and their men hanged in half the ports between the Rhine and the Biscay. By the time the Armada reached the English Channel they could raise only seven ships, the largest of them barely 120 tons, and the Ports themselves were passing, one by one, into decay. The eastward drift of sand and shingle into the Channel silted their entrances and filled their harbours.

In a sense the inherent strength of the Cinque Ports, however, had for generations prolonged the dependence of Britain on the Continent. It required the loss of Normandy to reconstitute it as an island. Rouen's surrender in 1204 is a climactic moment in the reign of John Lackland.

John's reaction to that disaster was swift and significant. The Rolls show a dramatic rise in naval expenditure in 1205. In 1206 Portsmouth was established as a naval base. By 1208 William of Wrotham had devised the nascent organization that became the Admiralty Office.

The dark malevolence of John was for seven hundred years the peculiar sport of historians. Modern judgement is more understanding.

He was warped, but he was a brilliant administrator, a man of vision, and he recognized the urgent requirement for a professional navy. Nine years after Rouen a small but effective British fleet existed and a primitive naval philosophy had evolved. Roger of Wendover, no friend of his, says respectfully of John that, threatened with invasion by Philip of France:

> . . . the king determined to engage his enemies at sea, to drown them before they landed, for he had a more powerful fleet than the French king.

In 1213 John justified his policy. Under William, Earl of Salisbury, a fleet of King's Ships and chartered vessels forced the channel which led to Damme on the Flanders coast. Much of Philip's invasion fleet was still drawn up on the beaches; the French knights and the majority of the men-at-arms were ashore, embroiled in the siege of Bruges. Salisbury's victory was absolute. 300 of Philip's ships were 'cast adrift', a hundred left in flames. It is the first significant offensive/defensive action in British naval history. Invasion was abandoned before the smoke of that burning died in the summer air.

In London John lodged claim to the 'Sovereignty of the Seas'. Possibly the claim was not yet valid; the strength of the fleet, the strength of Britain, fluctuated under successive kings, but after Damme Britain stood effectively in Ocean.

The ships that fought at Damme were largely open boats, at best half-decked, but already the King's Ships and some of the Cinque Ports' vessels represented major advances on the Norman types of William's time. Two causes are apparent: the first, John's recognition of the desperate need to provide an adequate defence against Europe; the second, the growing demand for sea-going merchantmen as a consequence of the loss of the short-sea trans-Channel trading routes.

Economically it was essential to work up the distant water trades to Aquitaine at the head of the Bay of Biscay, to the North Sea ports, to Scandinavia, the Baltic, and even as far as Iceland. The twin necessities are intimately connected: out of them and by way of the garnered experienced of the Crusades and the Mediterranean came the round-hulled, single-masted, decked ship with its rudder at the stern that, as the 'cogge', became the standard sea-going trading vessel of the Atlantic coast. The same ship, chartered and equipped for battle by the seasonal addition of summer castles – temporary fighting platforms erected at bow and stern – provided the basic warship. By the time Edward III challenged the fleet of Philip of Valois off Sluys in 1340 the type was perfected.

Contemporary accounts differ but Sluys is the second of the great offensive/defensive actions of the early navy. According to one account the French fleet had concentrated off Flanders under Hue Quieret for the invasion of England. Froissart says that the King 'put to sea with the intention of reaching Flanders to aid his brother-in-law the Count of Hainault'. It is agreed that Edward 'came sailing before Sluys', with Sir John Chandos singing, and broke the sea strength of France. The King wrote to the Black Prince 'under our secret seal':

> . . . trusting in our just quarrel, we entered the said harbour with the tide to attack our said enemies who . . . offered a very noble defence all that day and the following night, but God granted us the victory.

The Genoese galleys of Philip's fleet alone escaped under oars, and for a period England was supreme in the narrows of the Channel. With the battle of Les Espagnols-sur-Mer, Sluys is the terminal action of the ancient navies. After Sluys came the Gun.

There is no grand design in the development of British sea power. It is a history of *ad hoc* decisions, of reactions to external stimuli, of responses to internal pressures. It is individualistic, the consequence of royal enthusiasm and at least as often of royal apathy, of the advent at irregular intervals of commoners with sea sense and courtiers with foresight. It was never national, not even – despite the historians – at the climactic moment of the Spanish Armada. A king was to learn this at the cost of his throne and his head as late as the middle of the seventeenth century. It was always peripheral, the particular business of the coastal strip, of the ports, of the men who lived daily with the inescapable reality that Britain *was* an island.

But if there was no grand design, there was nevertheless a pattern. It has four component parts: the primary need to improve ships continuously against the competitive improvements of other nations; the permanent importance of extending the distant-water trades following the final severance from the Continent; the idiosyncratic acknowledgement by kings and rulers of its absolute necessities; and the basic and enduring practice of piracy. All four are inextricably interwoven, no single one is more important than another.

The adventitious appearance of the gun is typical of the pattern. It is just possible that it might have been used *at* Sluys; Edward is credited with using 'crakys of war' against the Scots as early as 1327, and three small bombards were certainly employed at Crecy seven years later. The adoption and the perfection of the gun at sea are, however, the special achievement of four of the English Henrys. Traditionally it is Henry IV, in the midst of his troubles with Parliament, who mounted 'three small iron guns' in the *Christofer of the Tower* in 1406. Henry V, pursuing his elaborate ambitions to re-establish a British presence on the Continent, built the *Grâce Dieu*, the first two-masted ship in the British register, and opened the way to an increased manoeuvrability and a consequential revolution in tactics. Henry VII, invading his own country with the help of the French, rebuilt a navy that had been virtually destroyed, grasped the essential importance of the guns that were available, and designed the four-masted *Regent*, the largest ship of his day, to carry 225 breech-loading serpentines. Henry VIII inherited from him seven sound ships, a base system carefully calculated to future needs, an administrative organization of ability, and a vision of sea power. Henry VIII made himself and England the masters of the gun.

The launching of the *Mary Rose* in 1513 marks the end of the medieval ship. 'The flower, I trow, of all ships that ever sailed', she was built with two gun decks. Watertight gun ports, invented possibly by James Baker, enabled her to carry 3,000-lb. brass curtals on her lower deck. She was the first ship of the English renaissance.

The inter-relationship between navies and trade is intricate. It is never simple to define the degree in which the requirement to protect trade creates the necessity for a navy. Equally it is difficult to assess precisely the extent to which the protection of trade promotes a navy and supplies thereby the requisite finance. Henry VII, bringing the Wars of the Roses – with their disastrous effects on trade – sharply to an end, 'could not endure to see trade sick' and used his ship developments to assist the new distant-water traffics. The *Mary of the Tower*, the *Sovereign*, and others of the King's Ships were, for example, chartered to the Levant trade. Henry VIII continued the custom and eased his problems of cost thereby. None the less, the available evidence is that Britain fell progressively behind in the development of the new oceanic commerce.

This is the central epoch of discovery, and despite engaging mythologies, Britain's part in the opening of the New World is all but negligible.

Henry VII had been on the throne barely three years when Bartholmeu Dias turned the southern extremity of Africa and opened the sea routes to the Indian Ocean. Immediately upon his return Christopher Columbus sent his brother Bartholomew to Henry's court to seek funds for an attempt at the western route and was, after hesitation, refused. Three years later, subsidized by Isabella, he reached the Americas. Six years later Vasco da Gama anchored off Calicut on the coast of India. From Britain only the Cabots, father and son, given leave (though not funds) by Henry 'to seeke out, discover and find whatsoever isles, countries . . . of the heathen and infidel, which before this time have been unknown', set forth in search of a belated share. The voyages of the Cabots, losing themselves in the cold complexities of the American North-West, were, in the long run, a disaster to the Bristol merchants. Made in search of trade, they lacked the ultimate drive of intellectual curiosity. Britain had had her four Henrys, she lacked a counterpart to The Navigator.

By the end of the reign of Henry VIII three components of the pattern had come plainly into relation. The integration of piracy remained to be accomplished. With Elizabeth it was the linstock which touched off the plenary explosion of British Sea Power.

It had, in fact, existed in Britain from a time beyond the memory of man. Possibly the Veneti brought it in its simplistic form of coastal raiding to the English coast before Caesar's raids, but classic piracy is more complex – basically it is a by-product of conventional trade. Britain required the passage of a constant traffic between the North Sea and the Biscay to elevate the practice of it to maturity. All the ports of the narrow waters along the Channel took part in it at intervals. The Cinque Ports, existing in a delicate balance between the legalities of cross-Channel trade, the tangled illegalities of cross-Channel raiding, and the endemic robbery of the east–west trade, were for a long period its principal exponents. Local rivalries among them and elsewhere engendered local feuding. Romney raided Yarmouth, Winchelsea and

Rye revenged themselves upon the Fowey Gallants, the Scillies became a pirate base against everybody, the coast of Scotland harassed all nations at the approaches to the Baltic, and Ireland then and later was a complex of sea brigandage.

But throughout medieval history piracy was also the established recruiting ground that provided men and ships for the build-up of the King's Fleet in time of war.

In the age of Elizabeth, expanding tumultuously with the increase of trade that followed the epoch of discovery, it swept across the Atlantic in the wake of opportunity to make itself a significant element in the power behind the Elizabethan throne.

John Hawkins was a pirate.

All the whitewashing appropriate to a national hero cannot disguise the fact that his first venture to the Guinea coast was a plain essay in hi-jacking. Attacking the Portuguese slavers of the coast, loading up with captured and some purchased slaves, he crossed the Atlantic to sell them by guile and force in the forbidden area of the Spanish conquests. Elizabeth chartered a Queen's Ship to him for his second voyage. Piracy became a joint stock operation. Hawkins was appointed, in process of time, Treasurer of the Queen's Navy.

Drake was a pirate.

The Spaniards called him the 'master thief of the known world.' Elizabeth arranged for him to be knighted for it on the deck of the *Golden Hind*, figuratively over holds crammed with loot taken in the voyage round the world from the Spaniards, with whom England was lawfully at peace, and Portugal, who was an ally.

Raleigh was intermittently a pirate, so was Frobisher, so were Gilbert and half a hundred others; but it was they who gathered and stiffened the fleet for Elizabeth when an angry Philip determined to end the intractable offence of England.

The strategy which defeated the *Armada Felisisima* – the appellation *Invincible* is a fabrication of propagandists – epitomizes the history of English Sea Power.

It opened with an offensive/defensive attack that matches John's determination 'to drown them before they landed'. Drake's entry into Cadiz harbour has the élan of Chandos's 'singing' before Sluys. Executed with all the ruthless techniques of the raids on the Spanish

Main, its plundering was in the high tradition of the Channel, but Drake's decision after it to base himself on Sagres athwart Philip's lines of communication was the result of an intellectual concept infinitely removed from even the most complex piracy. By the two achievements Philip's attempt was delayed for an invaluable year.

The tactics which defeated Medina Sidonia's plan of campaign when once the Armada had entered the Channel were a superlative combination of the technique of the raids on the Spanish treasure fleet and a mastery of the technological potentials of the modern gun.

The English ships were not of course smaller than Philip's, despite legend. They *were*, however, lower in silhouette; most of them lacked the high castles of the Spanish tradition with their top weight and their sea tenderness. They were ships built not for the easy waters of the West Indian trade out of the hurricane season but for the hard, cold winds of a northern island. And they were built, deliberately, to exploit the potentials of the longer range of the main English gun and the hard-bought experience of innumerable lawless attacks far from support.

The *Armada Felisisima* carried the medieval taint of Spanish military tradition. Its ships were designed to carry soldiers who fought by boarding. To board they had to grapple – and they could not grapple.

Once Howard gained the weather gauge off Plymouth, Medina Sidonia was, in fact, defeated. This was a sea battle of seamanship. Driven slowly up Channel at an average speed of less than three miles an hour, battered for nine days incessantly by guns that outranged their own, Sidonia's ships none the less made their rendezvous off Gravelines. But Britain was now decisively an island.

Parma, whose army was the vital element in Philip's plan to invade England – 'I think foul scorn of Parma' – was not ready for the sea crossing. Fire ships smoked out the Armada that was to be his convoy. The long haul round the island, the damaged state of the ships, exhaustion and disease among Sidonia's men, shortage of food and wine, gales and the coast of Ireland, shattered the *Armada Felisisima*, and in the glory and splendour of that victory the island of Britain became a World Power.

Like piracy a navy, even a victorious navy, requires trade to support it. The problem after the campaign of the Armada was to maintain a fleet of the necessary size and quality on the taxation possible from a traffic that was still in embryo.

Comparative figures in the areas opened by the explorations do not exist but a short-sea trade in the traditional area offers significant statistics. In 1590, two years after the Armada, 5,623 vessels moved through the Sound, the entrance to the Baltic. More than 2,800 of them were Dutch, fewer than 600 Scots and English. The Dutch largely had taken over the activity of the Hansa, they were the masters of the carrying trade. Almost half of Europe's total shipping in that year – 2,000,000 tons, by William Petty's estimate – belonged to them, less than one-eighth to England.

When Spain's seizure of Portugal deprived the Netherlands of the trans-shipment trade from Lisbon, the merchants of the Low Countries forced their way promptly into the Eastern trade itself. Britain, experimenting less purposefully, fell behind.

It is the defect of virgin queens, however splendid, that by definition they leave no heirs. The garrulous and dribbling James, succeeding Gloriana, neither expanded his heritage nor secured its continuing integrity.

Paradoxically, his reign saw the burgeoning of the tentative system of colonization that had begun under Elizabeth, but he provided neither encouragement enough to expand it nor a navy strong enough to sustain it.

His son Charles saw both necessities. It is possible that he saw them too clearly. Despite the tumultuous enthusiasm that greeted the victory over the Armada, Elizabeth, who had made England into a world power, had failed to establish her as an homogeneous sea power. England was still divided between the coastal fringe which lived by and recognized the importance of the sea, and the towns of the interior which still to a remarkable extent ignored the sea. The coastal fringe, living by it, paid for it. The view of Charles was that the defence of the country was, or should be, a charge upon the country as a whole. Ship Money was a rational attempt to spread the load.

The pattern inevitably grows blurred. Through the fifty years after the Armada royal enthusiasm had diminished. Ship improvement consequentially had faltered. The distant-water trades had failed to expand in parallel with those of Britain's greatest rivals, the Dutch. Of the four great components of the pattern only piracy remained, and profitable piracy had passed to hostile hands. Between 1609 and 1616 Algerian pirates took 500 English ships in the Channel. The Sallee Rovers attacked commerce in the Thames itself, the Dunkirkers raided Kent and Sussex with fleets of more than a score of privateers. The Newcastle coal trade was suspended. Scarborough and Yarmouth were 'miserably oppressed'. King's Lynn was idle, and in Ipswich seaborne trade ceased. London was reduced to fewer than ten deep-sea ships.

Charles's solution was just and equitable. Even with the reluctant Ship Money, however, even with the genius of designers like the Petts, who built the impeccable *Royal Sovereign*, he failed to re-establish a navy of a strength adequate to deal with his manifest necessities.

In the upshot Ship Money exploded the mine of the Great Rebellion. Certainly other causes existed, but the tax was the trigger, and it is one of the perverse twists of English history that it required a citizens' rebellion to re-establish a traditionally 'Royal' navy on an equitable basis, to give it the beginnings of a proper financial structure, to discipline it on contemporary principles, and to supply it not by way of seamen but by way of Generals at Sea, with strategic concepts and tactical doctrines that were to endure for two hundred years.

These things are in sum vital, yet Cromwell's principal contribution to the accelerating flow of British expansion is that slowly and, for so resolute a man, irresolutely, he drifted into war with the Dutch. For more than a quarter of a century the campaigns followed in irregular succession. The names of their commanders are like a roll call of history – Blake and Monk and Prince Rupert, Tromp and Evertzen and de Rüyter. The Dutch destroyed the Fleet in the Medway, the British burned 150 merchantmen in the Vlie Channel. Lowestoft, the Texel, the Gabbard, the Four Day Battle, the Foreland – the names of the sea fights are like a roll of drums.

Yet, considered dispassionately, the smoke of the Channel guns is scarcely more than a screen hiding from contemporary eyes the global importance of the new possessions which accrued to Britain across three oceans. When the third Dutch War ended in 1674 Britain had erected the sub-structure of imperialism.

Begun hesitantly on the island of Roanoke in Elizabeth's time, the process had expanded to take in Virginia and Newfoundland. The Eleutheran Adventurers had temporarily secured the Bahamas. Barbados, St Kitts, St Eustatius and Tobago were in British hands. Manhattan had surrendered. Half the world away the East India Company was securely established on the peninsula of India.

There is no accident in the fact that the greater part of the tentative colonizations were planted upon islands. Strategic potentials rather than territorial area are the significant characteristics of the acquisitions of this period. Islands are defensible; equally, they can be used to control sea routes.

Consolidation came later with the French wars. Half land battle, half sea war, the campaigns that began under the Grand Monarque in 1667 and, fostered by the genius of Colbert, lasted through his reign, continued under his variegated successors, and ended only at Waterloo.

They left Britain in all respects an Empire of the Sea.

The pattern at this point changes altogether – and permanently. Of the four main components, the royal influence had all but disappeared with the transfer of power from kings to Parliament. The semi-permanent neurosis over the maintenance of fleets had diminished with the expansion of overseas trade. Piracy had been driven out of the nearer waters, was at the point of abolition in the Mediterranean, and except for occasional romantic survivals like de Soto of the *Black Joke*, lingered only in Eastern waters.

This is the period of the quaintly designated Pax Britannica, the era in which possession of a dominant fleet made Britain the Policeman of the World, and peace spread wide across the seas. On land, oddly enough, it engendered almost continuous war, beginning with the Abyssinians and ending only with the Zulus, but that is beyond the scope of this book; it is enough to record that as the Pax Britannica extended its beneficence, Empire continued to accrue. In the Americas, in Africa, in the Middle East, the Indies, Australasia, the Pacific, even the Antarctic, possessions were claimed, colonies planted, dominions established. On this the sun, as the poets pointed out, could no longer set. The offshore island, in revolution, for a little dominated the world.

In parallel there was, inevitably, another revolution.

In 1788 on a tiny loch at Dalswinton in Scotland Patrick Miller, an Edinburgh banker, built a small, double-hulled boat, powered it with an engine and paddle-wheels designed by William Symington, and with a crew which included, improbably, the poet Burns, proved the feasibility of steam propulsion. Whereby he unlocked the Pandora's box of maritime technology. For almost half a century the Lords of the Admiralty splendidly defied the inevitable, refusing steam, rejecting the screw propeller, spurning the iron hull, scorning the submarines. But slowly, the French stealing a march at Kinburn with the first practical armour at sea, the Americans racing ahead with the *Monitor* and the *Merrimac*, the Fenians inspiring the earliest effective submarine, the Germans abruptly discovering the significance of the sea, and the Japanese emerging from medievalism as a fledged naval power, the last and greatest race in conventional aramaments got under way. In 1904 Admiral Fisher forced the *Dreadnought* – the world's first all-big-gun battleship – on a bruised and doubtful Admiralty, the Italians, the Germans, the Americans came up a bare ship's length behind, and in a tumult of threat and counter threat the Battle Fleet dominated the world.

Underneath the tumult, almost submerged in its clamour, two conditioning events took shape.

In the ineffable peace of a Cambridge laboratory Ernest Rutherford developed the theory of radio-activity, unregarded by the military mind.

In a quiet meadow under the Iron Age rampart of Dover Castle, watched probably by no more people than saw the first waves of the Flandrian Transgression break across the isthmus, Louis Blériot landed a precarious and unlovely aircraft at the end of the first flight across the Channel.

When they moved his machine on the following morning they found the cliff grass standing in its shape, outlined by the trampling of innumerable feet. With a rare inspiration they sank dressed stone in the outline for a memorial. It looks today like an oddly misshapen grave.

It is in fact the grave of the tradition of sea power.

After these things in the inexorable succession of progress came the bomber, the air fleet, the nuclear weapon and the missile.

Britain is no longer an island.

IWO JIMA

Britain was big as islands go. Malta is small. Iwo Jima is a meagre speck of land far out in the empty waste of the Pacific. Britain before 1944 had an immense and complex history. Malta had a record of 2,000 years of valour. Iwo Jima had no history at all.

It had, of course, been discovered. A Spanish expedition in the sixteenth century named four small islets – crests of one of the vast under-sea ranges of the Continental Drift that loop in swags down the Western Pacific – Santo Alessandro, Sulfur, Santo Agostino, and Arzibispo. Thereafter they were forgotten. The Japanese re-discovered them, probably at the end of the century, designated them the Kazan Retto, renamed three of the islands Kita-Iō-Jima, Iō-Jima and Minami-Iō-Jima, and forgot them again. An Admiralty chart of 1881 deals with the fourth islet, Arzibispo, in a blunt notation: 'Not seen by U.S. Ship *Alert*, 1880.'

In 1887, unremembered by the Powers even through the great years of the land grabbing, they were occupied, surreptitiously, by a small group of Japanese fishermen and sulphur miners. In 1941 at the outbreak of the Pacific War they were still forgotten.

Iwo Jima, the central island – the wartime spelling is now accepted – is less than five miles long, barely two and a half across its widest point. It had none of the traditional requirements of island strategy – no harbour, no ships based upon it, no hinterland. It could not conceivably grow crops to support even a frugal garrison. It lacked everything except position – it was approximately 640 miles from Tokyo and 635 miles from the Marianas.

When Saipan in the Marianas, the key point of the outer defences of Japan, fell in September 1944, Iwo became the halfway mark to Tokyo. What it lacked in the traditional strategy of the sea it was discovered to possess in the new strategy of the air. Potentially it was a major air base. With adequate aircraft, it could in Japanese hands defend the direct approaches to Tokyo. In American hands, it could supply a staging post and fighter cover for an all-out air offensive on Japan.

In September General Tadamachi Kuribayashi, commanding the sea area Bonin/Kazan, was ordered peremptorily by Imperial Headquarters to:

> . . . transform the central island into a fortress, secure the gradual depletion of enemy attack forces, and, even if the situation gets out of hand, defend a corner of the island to the death.

A fortnight later Admiral Nimitz, U.S. Naval Commander-in-Chief, Pacific Ocean Areas, received a directive which said in the terse and less accented language of Western strategic communication:

> Admiral Nimitz, after providing covering and support forces for the liberation of Luzon, will occupy one or more positions in the Bonin Volcano group, target date 20 January 1945.

With a singular economy of words these two documents spelled out the death sentence on 30,000 men.

Had it been possible to assault Iwo Jima in August 1944 immediately after the fall of Guam, it might have been rushed quite simply. Delays in MacArthur's campaign for Leyte Island prevented the release of ships even for the planned date of 20 January. Kuribayashi was given invaluable time.

Strategic bombing aimed at the interdiction of the airfields on Iwo and Chichi Jima had begun as early as June, long before the capture of the Marianas was complete. By the second week of December it had been built up to daily strikes, at the end of January round-the-clock bombing was initiated; simultaneously five heavy cruiser strikes were carried out.

In early February a bombardment force of eight old battleships, the aircraft carrier *Saratoga*, escort carriers, cruisers and destroyers left Saipan. I sailed with the U.S.S. *Tennessee*, one of the survivors of Pearl Harbour.

At dawn on 16 February we sighted Iwo Jima. In the early light it looked, with grim humour, like a swimming turtle. The summit of the cone of the 550-feet Suribachi volcano appeared to be separated from the main island as the head of a turtle in the water is separated from the carapace. Even in that early light we could see that it was a green island.

At 14,000 yards we opened fire with main armament, one component in a vast semi-circle of steel that closed slowly from the south-east on the inert land. The guns thundered, the sky was streaked with the scream of shells, the clouds crossed with patterns of aircraft. The world was full of intolerable sound.

Only Iwo Jima itself was silent.

All through the long day while the scrub on the island burned and the smoke and the dust of the explosions lifted in immense convolutions to the upper air, the bombardment continued.

When we withdrew, Iwo was still silent.

At dawn on the seventeenth Iwo was sere and yellowing. It was as if autumn had come in the night. Smoke still rose, fires still burned. It was difficult most of the time to see the outline of the island but we opened fire again and there was still no answer from the land.

We went on firing. It was impossible to believe that in the heart of that pulsation of dust and thunder men still lived and had their consciousness. They did. Somewhere about nine o'clock we saw the towering fountains of shell splashes in the sea between the ships. Immediately a cruiser hauled out of line, burning. Moments later a battleship was hit. Then, improbably as it had begun, the Japanese fire ceased. Iwo was inert again.

In the mid-morning twelve landing craft went past us. This was not the opening of the assault. This was a reconnaissance of underwater demolition teams, searching for beach obstacles. Instantly Iwo awoke. As the landing craft went in, batteries on either side of the landing beach unmasked. In minutes nine landing craft were out of action, the remainder damaged as, under a storm of counter battery fire, they pressed in far enough to make certain that there were no obstacles and withdrew. Iwo fell silent immediately. The bombardment proceeded.

In the dawn of 18 February Iwo was brown, slashed with purple and dark olive. Fires still burned. This day we blew the last of the ancient surface off the island. The flames of the napalm from the aircraft scorched ferociously across it. The shellholes interlocked. Nothing lived except the gulls that wheeled white against the blackness of the smoke above it. In stoicism Iwo endured.

There has been no other sea bombardment in history to match against it. When in the dawn of 19 February the Fifth Fleet joined us, Iwo was altogether ringed with ships and the crepitation of the 16-inch guns of the new battleships below the horizon beyond the island made the air shudder about us. Air strikes of a hundred aircraft at a time from the Fifth Fleet carriers roared in and piled their thunder upon the thunder of the ships. The world shook.

At 06.45 Admiral Turner made the signal: 'Land the Landing Force.'

Slowly *Tennessee* moved in to the central position off the landing beach, so close that our targets were almost under the muzzles of our guns. To seaward the amphtracks – the armoured amphibious vehicles – of the first assault wave formed up. We watched them from the bridge as they split round us and roared in the last few hundred yards. Somebody said softly: 'It's going to be a pushover.'

Iwo was still silent.

I remember hardly more of the first phase than the astonishing precision of that assault wave, the upturned faces of the men in the nearer amphtracks as they went by us, and the beetle-like movement as they hit the beach and we lifted the rolling barrage to let them through to the high ground above the sand. I do not remember hearing the opening of the Japanese guns. I do remember recognizing with a sudden painfulness that the first wave was in trouble, not with the Japanese but with the treacherous black volcanic sand of the landing beach that would not let the amphtracks in or, letting them in, trapped them once they had made a footing.

Into that turmoil drove the second wave. Amphtracks and landing craft collided. Vehicles died in the sand and were engulfed. The straight, narrow two-mile beach was a confusion of struggling vehicles and desperate men. And into it we were suddenly aware the Japanese were pouring an unbearable storm of fire.

On the slopes of the Suribachi volcano, in the black sand dunes above the beach, in the cliffs to the north-east, and in the high ground beyond, thousands of men had surfaced, coming up out of prepared openings, digging their way through where our shellfire had blocked them in. Gun positions were cleared of the debris piled over them. Multiple mortars were brought up from the tunnels of the sulphur mines and the hundreds of volcanic caves that Kuribayashi had cut and hollowed and joined and carved – in the time that had been given to him – into an island fortress as intricate as a *netsuke*. This was the explanation of the silence of Iwo Jima. In a seven-mile maze of tunnels 22,000 Japanese sat out the bombardment in immunity.

The silence was broken now, but very gradually the support fire of the ships began to exert a measure of control. Slowly the beach parties started to get order into chaos. At 09.44 the LCTs with the tanks went in, and slowly the leading vehicles clawed their way up the shifting sand to bring fire to bear on the nearer enemy. All the time fresh waves of marines went in. From the bridge of the *Tennessee* it had a macabre and terrible fascination; we were so close that through the powerful glasses we could see the faces of the men who fought, we were so far that it was impersonal. It was like an enormous and irrational spectacle. We had almost no sense of participation, no sense even of reality.

At dusk two divisions of marines were penned in an L-shaped strip of land that cut off the Suribachi from the main island, ran across the end of Airfield No. 1, and stretched along the beach. It was barely half a square mile in area. It had cost 2,400 casualties.

All through the night they clung to it. At dawn they began to inch their way forward. A ten-foot ridge of dune represented a tactical victory, twenty feet of upward slope a triumph. All through the day they fought while the ships pounded the mainland and the Suribachi smoked as if a new eruption had begun.

It set a pattern for the days to come. Fifteen days were needed to take a third of the island and to establish a footing on Airfield No. 2, a thousand yards away. It took five assaults to carry the Suribachi five days after the landing. We saw the first flag raised there as we changed stations in mid-morning. The flag that the American correspondent Rosenthal photographed an hour later was brought from a wrecked L.C.T. It made perhaps the greatest picture of the Pacific War.

No help came to Kuribayashi. The numbers and the disposition of the U.S. Navy made effective relief impossible. Even the vaunted kamikaze failed. Two attempts were made. As we withdrew for the night of 21 February, the aircraft carrier *Saratoga* was attacked – one kamikaze crashed into her, bombs hit her and she passed out of sight, blazing. None the less, she survived. The Japanese Navy itself made one attempt against the invasion fleet with the kaiten, the one-man suicide torpedo. Two of the submarines that carried it were destroyed. No damage was done.

On 16 March Iwo Jima was declared 'Secure'. By 26 March, when the operation was officially deemed complete, the U.S. Marines had captured 216 prisoners – the great majority of them unconscious or helpless at the moment of capture. 20,703 dead had been counted. There were still men underground; for a month the rat hunt continued: 1,602 more dead were added. The Emperor had defined their task:

> You will defend a corner of the island to the death.

They had.

It cost America 7,000 dead and more than 20,000 wounded.

Refuge

CAPRI

Peace on islands, as upon the mainlands, is the obverse of the medal of war. Refuge is an aspect of peace. If the proposition put forward in the opening chapter of this book is justified – that an escaping tribesman on a forked log first discovered the immunities of the offshore islet – refuge is one of the primary aspects of islands.

Leto found on Delos a refuge from the wrath of Hera. The Phoenicians – from Tyre to Cadiz – sought in the offshore islands a refuge from the hostilities of the beginnings of trade. Anchorites and hermits, monks, priors and abbots through the Middle Ages made refuges for Christianity in the islets of the Western world. Pirates, sea-robbers, idealists and romanticists of every conceivable variety and eccentricity used them. It is a process as old as time, older certainly than history and patterned with failure and success. It continues in our day. The nudists of the Île du Levant are now fighting a slow and painful battle in the course of which they have been driven to the edges of their island to herd naked like seals on the harsh rocks and in the crescents of minuscule beaches by the power of the French missile industry – which has created a phallic symbol more potent than their own.

Refuge is a necessity of emperors no less than of the common man. Capri was the refuge of an emperor.

It has a complex story. It is appropriate, therefore, that its origins are complex. Neolithic man lived there; his remains have been found in the Grotto delle Felici. It was involved in the Euboean invasion which settled Pithekoussai and Cumae to the north in 750 B.C. It may have been utilized briefly by the Phoenicians – race memory endures in the Phoenician Stairs. It must have been used by the Rhodians when they founded the colony of Parthenope, which became Naples. Its inhabitants were basically of Greek descent when Augustus exchanged it with the Neapolitians for Ischia thirty years before the birth of Christ.

Augustus had no need to use it as a refuge. He exchanged it for fun purposes. He was the counterpart of George IV across the centuries, the enthusiast of the seaside holiday.

Tiberius developed a need for it.

As a boy he must have known the island. He was the Emperor's stepson, ceded with his mother to Augustus eleven years before. Almost certainly he would have gone there with Augustus, or at least to Baiae – the Cap d'Antibes of the Campania – and Capri was visible from the terraces of Baiae.

But Tiberius, born to great events (his mother and her maids hatched an egg in their warm hands before his birth and it was a cockerel with a potent comb) left the court circle early. At twenty he was sent to the army, to the campaign against the Cantabrians; for almost sixteen years, except for brief winter visits, he was away from Rome.

He returned to the city in 7 B.C. to a full Triumph, not perhaps a soldier of genius but a leader of unremitting determination. Then, 'in the prime of life, in excellent health, and at the height of his career', according to Tacitus, he withdrew from state affairs – and from Rome.

Augustus attempted to hold him, but he began a hunger strike and Augustus surrendered. Tiberius was permitted to sail to Rhodes, where he surrounded himself with Greek scholars, abandoned the great world, and was almost abandoned by the Emperor.

After seven quiet years he was permitted by Augustus to return, and immured himself in a house in the Gardens of Maecenas while the fatal complexities of Imperial politics revolved around him.

He emerged the Emperor's formally adopted son – Emperor designate so far as anything was ever certain in the Roman succession – and went back to the wars. For almost six years he fought victoriously. He was on his way back to the field from a brief period in Rome when he was recalled to the death bed of Augustus.

He was fifty-six at his accession. Of the thirty-six years since his first campaign, he had spent fewer than seven – and these at widely separated intervals – in the capital. It is difficult to escape the assumption that he hated Rome. He could at least have taken a campaign off; he need not have gone to Rhodes; and there is at any rate clear logic in the conclusion that he decided to evade Rome's importunities.

Perhaps it was not the city that he hated – though it is curious that almost alone among the Emperors he added nothing of importance to her. Possibly what he hated was Rome the demanding, ferocious, political entity; the Cosa Nostra politicians of the Senate, the corruption of finance, the unlimited intrigue, the plutocratic society and the stinking slums, and the insensate ambitions of the women.

Perhaps it was the women who worried him most. Vipsania, his first wife, he had loved, but Augustus forced him to divorce her after she had borne him her first son, Drusus, and while she was pregnant with another. He was compelled to marry the Emperor's daughter Livia – a whore notorious even by the standards of patrician Rome.

Tacitus claims that it was Livia's unfaithfulness that drove him from Rome originally. Tiberius himself said that he went to Rhodes to leave the succession free for Gaius and Lucius. Both these things were an integral part of the complex of the Rome that he had to escape.

The throne and its immediate necessities blocked the way, and the first nine years of his reign were singularly dull. He was conscientious, as painstaking as he had been as a general; but he was also, beyond question from the Roman point of view, a bore. He lowered the pay of actors and limited the number of gladiators at festivals, he refused honours and regulated the price of fish. He issued an edict against promiscuous kissing. He was unostentatious, unforthcoming, unsociable – and, understandably, unloved.

In the thirteenth year of his reign he rode out of Rome – and never went back.

Tacitus says that Sejanus, the ambitious commander of the guard who had made himself indispensable to Tiberius, engineered this to get the Emperor away from other influences in Rome and to further his personal aspirations. It seems at least improbable; the evidence is much stronger that Tiberius used Sejanus.

Suetonius declares that Tiberius employed the need to dedicate a temple in Campania as a pretext to get away from Rome, 'crossed over to the Isle of Capreae which fascinated him by having only one small landing beach', and let all affairs of state slide.

Emperors had small need for 'pretexts' and, far from letting affairs of state slide, Tiberius seems to have exercised from Capri an unremitting remote control on the affairs of the capital by a system of signals through a carefully controlled post on the Punta Campanella. In due course indeed he even organized from there the elimination of the indispensable Sejanus – now Consul and 'the partner of my labours'. After which, for six years he stayed on in Capri and the Empire did reasonably well.

Why did he choose Capri?

In the first place, he owned it. He may have had boyish memories of its possibilities. He knew its advantages: it was far enough from Rome to be clear of the defects of the city, close enough for the efficient Roman post to keep him in immediate touch and it was difficult of access even for the importunate.

Both Tacitus and Suetonius claim that he went there because it was defensible. It wasn't. It had in fact *two* small landing places – the Marina Grande of today, to be used except in northerly winds; the Marina Piccolo, to be used when the north wind blew. Nor were the cliffs, as Suetonius claims, unscaleable. Lamarque, with a small French force, scaled them at night 2,000 years later and captured the British Governor, Sir Hudson Lowe – a humiliation that was to have its effect on the exile of another emperor on another island, St Helena.

Nor is there convincing evidence that Tiberius ever went in fear of his life after Rhodes. What he wanted, very simply, was a place where Rome could not follow him. Capri supplied it.

What were the consequences of his withdrawal?

To Rome itself, as far as one can judge, remarkably few. The Senate presumably missed an individual to whom it could be obsequious in its recurrent periods of masochism and whom it could, on occasion, defy to relieve its aggressions. The people in all probability failed to realize the significance of his absence until they had grown used to the fact of it. He was, after all, scarcely a sporting and in no sense a gregarious emperor; they must have remembered what he had done to the actors and the gladiators.

As far as the Empire was concerned there were minor disasters and major incompetences at intervals, which can possibly be attributed to his absence – the Parthians overran Armenia, the Dacians, Moesia; and the Germans invaded Gaul – but the average level of power appears to have been maintained.

It is the stated consequences of the withdrawal to Tiberius himself that have fascinated scholars through the ages. Tacitus is splendidly blunt about it.

> Here on Capreae, in twelve spacious, separately named villas, Tiberius . . . spent the time in secret orgies, or idle malevolent thoughts.

Suetonius is even blunter.

> Having found seclusion at last, and no longer feeling himself under public scrutiny, he rapidly succumbed to all the vicious passions which he had for a long time tried, not very successfully, to disguise.

The reputation of Capri is based on those passages, and on their subsequent elaboration. On the superb contrasts of sex, sadism and sunshine 150 years of whispering campaigns have built an edifice of delicious and titillating scandal for the modern tourist.

Tacitus reached his pinnacle with his descriptions of the first, odd excursion that Tiberius made to the outer limits of Rome in 32. He had gone then as far as Caesar's gardens and had turned back down the Tiber. The reason, according to Michael Grant's translation, was that:

> . . . his criminal lusts shamed him. Their uncontrollable activity was worthy of an oriental tyrant. Free-born children were his victims. He was fascinated by beauty, innocence, and aristocratic birth. New names for types of perversion were invented. Slaves were charged to locate and procure his requirements. They rewarded compliance, over-bore reluctance with menaces, and – if resisted by parents or relations – kidnapped their victims, and violated them on their own account. It was like the sack of a captured city.

Suetonius is even more specific. Robert Graves's translation says:

> On retiring to Capri he made himself a private sporting-house, where sexual extravagances were practised for his secret pleasure. Bevies of girls and young men, whom he had collected from all over the Empire as adepts in unnatural practices, and known as *spintriae*, would perform before him in groups of three, to excite his waning passions. . . . He furthermore devised little nooks of lechery in the woods and glades of

> the island, and had boys and girls dressed up as Pans and nymphs posted in front of caverns or grottoes; so that the island was now openly and generally called 'Caprineum', because of his goatish antics.

It sounds remarkably like Swinging London, or Swinging New York, or even Swung San Francisco. Nothing is missing – permissive homosexualism, group sex, voyeurism, pornographic pictures, nudie shows, even topless and, so to speak, bottomless waitresses. Only the blue cinema is absent. It had not yet been invented.

How much of this is true?

The eminence of Tacitus is unchallengeable. He had a singular background of official position and a breadth of personal knowledge that was measureless. But he was a republican, a believer in the ancient virtues. Brought up under the excesses of Nero and, on his own admission, a participant in the reign of terror of Domitian, he had a guilt complex to assuage. He was happy to believe the worst of emperors.

The popularity of Suetonius need not be argued. He had the universal appeal of a Sunday-paper journalist of 2,000 years later – an infinite capacity for sex and sensationalism. He was enthusiastic about dirt.

Neither man lived in the reign of Tiberius. Tacitus wrote the *Annales* in Trajan's time, seventy-five years after Tiberius's death. It was *not* discussed with eye-witnesses. Suetonius wrote *The Twelve Caesars* about 138, presumably after he had been dismissed as secretary to the Emperior Hadrian for frivolling about with Hadrian's wife. Both men used traditional sources: state papers, the *acta senata*, the *acta diurnis*, the Imperial archives. Both men must have been familiar with – though only Tacitus directly acknowledges it – the scandalous and vanished memoirs of the younger Agrippina.

This is the heart of the matter. Neither of the Agrippinas was exactly lovable, the younger was complex and a total hellion. She herself had no reverence for emperors – after all, she poisoned one. She would not have spared Tiberius in an account of her own scabrous life. The obvious parallels in the aspects of the accusations of both Tacitus and Suetonius share a remarkable feminine timbre – abnormality visualized from an abnormal woman's viewpoint.

That Tiberius was ruthless on occasion is, of course, true enough. A Roman emperor was so by definition. A Roman emperor's family was, by similar definition, his most vulnerable point. His principal advisers provided the next. The destruction of Sejanus and his associates was inevitable – emperors needed to anticipate treachery to live. But that Tiberius was forced to hide in Capri because of 'monstrous cruelties' is absurd. The legend of endless treason trials and incessant executions has long been mathematically exploded. That the Capriotes will continue to point out the Salto di Tiberio with crepuscular enthusiasm is, of course, inevitable. After all, Tiberius, living on Rhodes on borrowed time, was accused of throwing people over cliffs there too. It proves nothing except the tourist importance of a horror legend.

That Tiberius indulged in sexual eccentricities is probably true enough. It was, after all, a custom of the time. Augustus, who preceded him, harboured in old age, according to Suetonius, 'a passion for deflowering girls – who were collected for him from every quarter, even by his wife!' Nero, who succeeded him, became a comprehensive byword. Rome accepted both impassively for years. That Tiberius was compelled because of such diversions to hide his guilt on Capri is a manifest nonsense.

Tacitus first and Suetonius after him pile accusation upon defamation. In the end it is too much.

Tiberius was sixty-nine when he settled on Capri, an old man, worn with much campaigning, tall, gaunt, hook-nosed and bent. The process of senile deterioration may reasonably have begun; on the other hand, he may have suffered from an irritation of the prostate gland or carried through his life the traumatic consequences of mammary deprivation, or he might have experienced that splendid omnium gatherum of the psychiatrists, childhood shock. There was, if either of the authors is to be believed, something singularly nasty in almost every woodshed in Rome for him to see. What he could not possibly have had, however, was the necessary stamina.

By the time Sejanus was dealt with – the time, according to Suetonius, when his perversions were finally revealed – he was seventy-three. Between them the two authors attribute to him almost every known sexual deviation. It would have required an Olympian sexual athlete to deal with all of them – or to want to. Some of those Pans and nymphs – all dressed up and waiting – must have suffered from frustration, some of his 'minnows' not been satisfied. Could he really have endured for ten years those threesomes, those pornographic playlets?

London audiences after barely four or five years of the permissive theatre are beginning to acknowledge boredom.

If neither fear nor politics nor sex, then, kept him on Capri, what, apart from his attitude to Rome, did?

The paths to the Villa Io – the Romans seem never to have built a road – move up out of the town of Capri with a purposeful serenity.

Capri town is a three-dimensional opera set. Its white walls and its flowers are real and scented, and the paths – they are the original litter paths up which the Emperor was carried – follow an easy contour so that he would not have been unduly tilted. They pass today through banks of plumbago, cascades of morning glory, showers of bougainvillaea, and they come in time to the soughing pine trees of the cliff top.

The Villa Io – the Villa Jovis – the Palazzo Timberio – crowns the summit of the Capo. It is not a palace. It was never more than a large country house, it would scarcely have provided a pavilion for Hadrian's villa. It was not ostentatious – though it crowned a mountain top, it was designed with most of its bulk behind the crest and it cannot have been more conspicuous than the Greek chapels which crowned every island summit in the Aegean in after years or the castles that fringe the coast of Spain. It was not vulgar.

A thousand feet below the terrace that fronts its principal rooms, the sea lies in an authentic azure. The Punta Campanella, the great rostrum of the Sorrentine peninsula, cleaves it like a trireme, the recession of the facets of the cliff marches to the Gulf of Salerno. To the east the sea reaches in to where Herculaneum stood, and the terrace to which Tiberius went, down a sloping walk splendid with tesserae, opens on the whole vast curve of the Gulf of Naples.

It is one of the breathless viewpoints of the world. Every man makes his own judgements – Sunion in a hazy dawn, the Himalayas from Darjeeling, Hawaii from the north-east, Corcovado from a circling aircraft, Angkor Vat against a thunderstorm, Table Mountain from an open boat – the list is endless. Italy from Capri has with them a quality of timeless beauty, the singular perfection of detachment.

Was that what Tiberius sought?

He spent nine months in the Villa Io itself after the extirpation of Sejanus. A long time in a fading life, a long time in the reign of an emperor who had just destroyed the remote control apparatus of Empire that he had created.

What held him?

Not, certainly, solitude for murderous impulses nor alsatia for sexual abnormality. Either – both – would have been easier in Rome itself, as unregarded as, subsequently, Nero proved them.

What, then, could it have offered this grim, withdrawn old man apart from the fact that it was not Rome? The Villa Io possessed two things only – beauty and tranquillity.

Even Tacitus acknowledges the first of these. In what is almost certainly the first essay in the gamesmanship of tourism, what might reasonably be called the 'I was there *before* . . .' syndrome, hc says:

> The bay it overlooks was exceptionally lovely – until Vesuvius' eruption transformed the landscape.

HADRIAN'S ISLET

This island lies peacefully on a hilltop thirty miles from the Mediterranean.

Hadrian built it.

Italica to the north of Seville where Hadrian was born is sixty miles from the nearest water – the Atlantic. His personal enthusiasm for the sea itself appears to have been negligible. He did visit the island of Britain, that at least is documented; he examined the islands of the Aegean – basically for architectural and literary purposes; he ferried over to Sicily; but he had no intimate and direct links with islands as islands. He was not even interested in the islets off the coast of Campania, that playground of escapist Rome, though the record shows that he visited Baiae at least once – he died there!

Why then did he build it?

The sea itself he used because it existed and because at times there was no alternative. He was the Travelling Emperor, passing his devious way around and across the Empire like a weaver's shuttle, and it needs to be remembered that Rome faced two oceans then – the Atlantic and the German; that for a brief moment with Trajan's race to the Gulf of Persia she actually faced three; and that at the height of her power she entirely surrounded the Middle Sea. Hadrian had to use water because it supplied a crucial element in his communications network. He seems to have had little liking for it, and little to do with ships except as an Imperial passenger.

He was indeed at the far side of the Mediterranean when he became Emperor – at Antioch as Governor of Syria and Commander-in-Chief of the Eastern Army – and the background of his accession still has unresolved problems. His claim was that on 9 August, A.D. 117, information reached him that Trajan had adopted him as his son and by extension as his heir apparent. More certainly on 11 August he received the news of Trajan's death.

Rome, with her traditional suspiciousness, found the gap a little too apt. Hadrian, secure in the backing of the Senate and with the army in his hand, waited at Antioch for the dust to settle.

He waited equally to put into motion the beginnings of the tremendous process of securing the frontiers of the Empire that is the monument of his reign. In his last years Trajan had over-reached even the strength of Rome. With a frenetic emulation of Alexander, he had forced his way to the beaches of the Indian Ocean – railed at his age, 'Ah, had I Alexander's years!' – and pulled back, realizing late that he had gone too far. Hadrian, judging the dangers of over-extension yet more vividly, handed back Mesopotamia, Assyria and Armenia, disregarded the protests of Rome, and sat on at his power point, Antioch, with his army ready. Rome subsided, muttering suspiciously about the succession.

The record is, for a little, obscure but Moesia revolted and Hadrian made his way there, considered abandoning it as well and withdrawing to a firm frontier on the Danube bank, decided against this and, deftly reconquering the province, proceeded imperially to Rome.

He spent two years there – it is difficult to make it more. In this time he placated the people, secured a still closer grip on the Senate, developed his intentions about the Roman frontiers, determined his immediate objectives – and drew up his first design for a country Villa.

It is important at this point to recognize that there is a limitless difference between the Roman and the Western concept of a Villa. The word itself is held to be a diminutive of *vicus*, which meant a village, sometimes even a small town. It did *not* mean a modest country house, still less the hyperboles of a Victorian suburb.

Precisely what Hadrian had in mind at the outset is not in any real sense clear. He was much criticized even before work began for his choice of site, principally because it was not Tivoli on the steep hill above. Tivoli was the in-spot of the Roman jet-set of the 120s. In spite of the persistent folk memories of modern guide books the site was, and is, pleasant enough even for an Emperor – a broad and easy ridge above the valley of the Aniene, then, probably as now, surrounded with olive groves. There was one existing building, small and, from the fragments that have been identified, it dated back to the time of Sulla. Apart from that the site was spacious, the slopes moderate, and the contours appropriate for large projects. At Tivoli large sites tended to be on edge, and the neighbours were much too close.

Work began, according to modern dating, in A.D. 118 and was divided into three main periods. The first of these covered the seven years to A.D. 125. Detailed plans for this first phase must have been completed with remarkable speed to cover the principal buildings, for Hadrian left Rome early in 121 on the first of his gigantic inspections of the Empire. The initial stage of this took him through Gaul to the Frankfurt area and the *Limes Germanicus*, the north-eastern frontier of Rome. The actual frontier line was adjusted at necessary points; the ditch that marked it deepened everywhere; a palisade of split tree trunks set up; and at the Saalburg, the best known viewpoint of the *limes*, the weak Domitian fortlet was replaced by a cohort fort, properly garrisoned, as part of a general plan to improve the reaction time of the defence.

From Frankfurt he moved down the Main and the Rhine to the existing position of The Hague, established the market town of Forum Hadriani (Voorburg), inspected the military dispositions at the Rhine mouths and crossed to Britain.

Hadrian's Wall, striding across the neck of Britain from the Tyne to the Solway, is one of the great masterpieces of frontier building, the most splendid monument standing to the military art of Rome. Planned before Hadrian reached Britain by Falco, the governor he had sent to put guts into the hurried defence improvisations begun in Trajan's time, it was already rising rapidly across the northern moors.

The threat of the tribes was growing fast. Roman losses in the tangled and obscure fighting in the Lowland country had been heavy. Falco's plan, a quantum leap forward from any existing frontier system in Europe, provided for a wall along the high crags from sea to sea, turrets to guard it, and gateways and fortlets at every Roman mile to provide its infantry garrison with sally ports for an offensive/defensive system. The whole was backed by the rehabilitated forts of the ancient Agricolan defence line in the valley to the south.

When Hadrian reached it – '*ambulare per Britannos*', as his more offensive critics in Rome phrased it – it was perhaps five courses of stonework high. Reassessing with ruthless brilliance the strength of the Northern threat, he stopped the work overnight and freshly designed the plan to employ horsemen in a vastly more powerful form of the offensive/defensive. Cavalry forts were superimposed across the line of the Wall where it had already been built, the new stonework was torn down to accommodate them, and the great fighting ditch was abandoned. Hadrian then rode south to Londinium and Rutupiae and the Channel crossing. Before the summer was over he was in The Province.

Whatever he may have begun in Rome before he left, in Germany he had built a palisade, in Britain a Wall. Now in Provence, at Apt, he built a memorial to Borysthenes, his favourite hunter which he may have ridden to death on the journey south, and at Nîmes he erected a temple to Plotina, Trajan's wise and courageous wife, who had supported him in the difficult days of the accession and may very possibly have 'adjusted' the date of his adoption – it was worth at least a temple.

The creative urge is one of the most dangerous of all diseases. It is cumulative. Despite it, Hadrian held to the problem of the frontiers as his primary task. He went from Provence to his native Spain (almost certainly riding), crossed it inspecting military commands, hurried from there over the Straits of Gibraltar to Mauretania on news that rebellion had broken out, and settled a possible serious breach of the south-west border with a brisk campaign. By 123 he was on the Parthian frontier at the extreme eastern limits of the Mediterranean dispelling, in consultation with its king, the dangerous threat of full-scale Eastern war. By the spring of 124, with this threat in its turn resolved, he was in the Aegean, almost certainly with the bright flowers of Delos. In the autumn he was in Athens.

The Hellenization of Hadrian is a fascinating study. Perhaps it would be unfair to say that he became more Greek than the Greeks, but evidence is apparent. Other Roman Emperors had built in Athens, Nero had sung there and damned for ever Greek pretensions to musical taste; Hadrian immersed himself instead in its spirit. He was initiated into the Eleusinian Mysteries, he presided at the Greater Dionysia, and he launched into an immense programme of what an age less conscious of inspiration calls public works but which to Hadrian were clearly private satisfactions. He completed the vast temple of the Olympian Zeus, begun by Hippias and Hipparchos 650 years before and postponed in despair by successive generations of Greeks. He built the triumphal arch close to it marking the gateway from the old city to Novae Athena, his new Athens; he erected the wall that protected it; the elaborate Stoa of Hadrian with its library to the north of the Acropolis; the Temple of Hera; the Athenian Pantheon; a gymnasium; an aqueduct – all appear to have been started within a single year. By mid-winter in 126 he was back in Rome, a five-year hegira ended, the frontiers – and Athens – secure.

The first phase of his Villa was approaching completion.

No precise time-table for its construction exists. Approximate dates have, however, been established by Herbert Bloch's brilliant identification of the stamps on the bricks used in the work. Some further assistance is derived from the alignment of groups of buildings, from the names attributed to them, and from stylistic changes.

Not all these are reliable. It is well to start by accepting the simple fact that the Villa, with its four main building lines established in a reckless disregard of one another, is, as a whole, a planner's nightmare. If any of the drains ever worked, it was a miracle. On the evidence of the bricks, however, the first group to be completed included the northern portion of the Palace, the great courtyards above the pleasant valley – which Hadrian is said to have rechristened the Vale of Tempe – and which comprised the Courtyard of the Libraries, the Stadium, the suite of baths to the west, and the circular construction known as the Teatro Marittimo.

The second group includes the superb Piazza d'Oro – the final section of the Palace – the vast Stoa Poikile, the small baths, the large baths, and the Accademia on the ridge overlooking the valley of the Risicoli. It was probably completed in 133.

The third group involved the excavation of the shallow valley of the streamlet between the ridges to provide a setting for the long pond known as the Canopus, the remarkable hemicycle of the Serapeum, and the lesser buildings along the sides. It was probably incomplete at Hadrian's death in 138 and it may have been added to subsequently.

The alignment of the first group was determined at the start by the ridge above the Vale of Tempe. If the sun bath in the thermae below it was indeed built simultaneously, it has an alignment of its own, very oddly attached to an extreme corner of the main block of the Palace; it may, however, be an afterthought. The Teatro Marittimo is altogether on its own.

The alignment of the second group separates into two parts. The

Accademia, like the Palace, follows the general line of its own ridge, this time overlooking the Valle di Risicoli. The Poikile, on the other hand, runs positively east and west, with the great ranges of buildings adjoining it observing a severe rectangular agreement.

The alignment of the third group is dictated by the excavation of the valley between the ridges. Despite the fact that the two main sets of baths are attributed to the second phase, they follow the third line.

The names do not help so much as add splendidly to the general confusion.

The trouble was started 150 years after Hadrian's death by Spartianus, and has been accepted slavishly since. It was Spartianus who created the impression of a species of Roman 'Empire Exhibition', a collation of quaint architectures, collected by Hadrian in his travels and Romanized by him in retrospect.

> To its different parts he assigned the names of celebrated buildings and localities such as the Lyceum, the Academy, the Prytaneum, the Canopus, the Stoa Poikile. . . .

Spartianus was largely inaccurate anyway, but the modern assignment of the names has virtually no validity at all. It dates mainly from the last years of the fifteenth century when the Borgia Pope, Alexander VI, loosed on the ruins the vulture horde of looting cardinals that ravaged them for the next two centuries. The names seem to have been applied by them at guesswork. The Stadium was not a stadium but a nymphaeum, the Libraries were not libraries but open-air triclinia. The Canopus was at least a canal, with one aggrieved crocodile; but the Poikile was not a 'painted porch', it was a vast enclosed garden with a superbly proportioned pond and its immense covered walks were in sum about a score of times the length of the quiet gossiping-place of the Stoics beside the market of Athens.

Hadrian was not a plagiarist, he was a creative artist, influenced unquestionably by what he had seen but searching always for his own concepts of perfection – and sometimes finding them. How much of the design was his is not always clear (he must at least have had somebody to do his quantity surveying for him), but the inspiration was beyond question his.

Changes in style are therefore at once less probable and more informative. Hadrian was responsible for the climax of the hundred years of Imperial building which had begun with Nero, the years in which Rome broke away from plain engineering and tentative experiment with the Greek, and developed immense new methods of enclosing space and new ideas about external vistas.

The first phase of the building displays this clearly enough. The second demonstrates a new appreciation of pure Greek acquired during his active stay in Athens, but displaying even more vividly his mastery of space and perspective, as in the Poikile with its ultimate prospect down the valley of the Tiber past Rome towards the sea.

The third shows almost nothing of Egypt save perhaps a reminiscence of Antinoöpolis (the city he built in memory of the boy friend Antinous) in the lay-out of the Canopus with its one disgruntled crocodile and its highly variegated collection of statues. Even the possibility that he worshipped the bull god Serapis in what is called the Serapeum is lacking in proof, and it is argued that the final gallery of the Serapeum was in fact meant to represent the Hall of the Oracles.

It seems most reasonable to accept the Villa as a masterful and idiosyncratic presentation of the Roman view of Greece.

All of it – except for one building.

The Teatro Marittimo is not Greek, it is individual – an essay in private perfection for an entirely personal requirement.

Externally it must have looked astonishingly like a twentieth century oil tank, in brick. There was scant exterior ornament, though Count Fede – the first effective conservationist of the vast ruin – found a red marble faun in an outside niche of it. Its alignment is eccentric, altogether different, with its bearing of north-north-west by half-west, from anything in its neighbourhood. Its site is even more remarkable. The eastern sector of its exterior wall rubs hard against the western wall of the Courtyard of the Libraries. The entire Courtyard is set back roughly twenty metres from the line of the west wall of the Palace proper to accommodate it. The doorway of its vestibule faces away from the Palace and independently towards the apex of the ridge. The entrances to it from the Courtyard of the Libraries are inconvenient, architecturally indefensible, and clearly afterthoughts; the entrances from the Hall of the Philosophers, worse. Everything about it suggests that it was built before the Palace, in isolation, and that it was engulfed in the surge of construction that eventually followed it.

The answer to almost all the questions it raises lies in its interior. It cannot, in its intricate perfection, have been anything less than the individual preserve of Hadrian himself.

The vestibule is unimportant. It had space for a guard of perhaps four men, not more. It was entirely outside the drum, but it opened into a circular portico. Forty delicate, un-fluted Ionic columns supported the portico's circular roof. The curved wall of its pavement enclosed a circular sea. The sea surrounded a circular island. The island supported an exquisite composition of small pavilions as perfect as the irregular petals of an antique rose. At the heart of it, reversed curves enclosed a central fountain as vital as the pistil of a flower. The whole was forty-two and a half metres in diameter – roughly 140 feet – and the island itself less than eighty.

Nothing indicates a 'Teatro'. Only the surviving part of the gay and delicate frieze of the entablature of the central spaces, with its engaging tritons, its sea horses and its nereids justifies the Marittimo – that and, of course, its circular sea. Cardinal Farnese began its pillage, Cardinal Ippolito d'Este completed it.

Circles within circles are one of the sign manuals of Hadrian. The dome of the Pantheon, for example, is a calculation of concentric circles that has outlasted 2,000 years. His own tomb is a composition of circles that became eventually the keep of Rome. The shape of this island is explained simply enough. What was its purpose?

Security an Emperor could command. It seems to have had little enough importance at the Villa. If the quarters of the Praetorian Guard were really in the Hundred Small Rooms at the base of the Poikile, it must have meant that Hadrian had no fear – the position was tactically ridiculous. Privacy an Emperor could obtain at any time even in the great halls of his palaces provided that he accepted the customary presence of elements of his entourage. Quiet an Emperor could decree (except perhaps for the noise of the doves in the pine trees) if he backed the decree with threat.

The one thing impossible to Emperors in their official lives was to escape from the human eye.

Only those who have moved with the endless processions of a royal progress can evaluate the effect of the impact of the human eye. It holds everything – welcome and rejection, love and hatred, admiration and contempt and vengefulness. It is impossible to escape it. In the end it is bruising. Hadrian built his miniature island not as Marie Antoinette built her dairy, to be the cynosure of eyes, but to escape altogether from eyes.

Everything suggests it. The high, blank outside walls, the separation of the guard from the interior, the empty portico of the pavement on which no one could loiter unseen, the silent water of the mini-sea, the two little floating bridges that could be pulled back against the curve of the island with silken cords to cut it off even from its outer circle.

Inside with the bridges drawn was security, privacy, silence, and the absolute absence of the baleful stare. It is an island of refuge from the human eye.

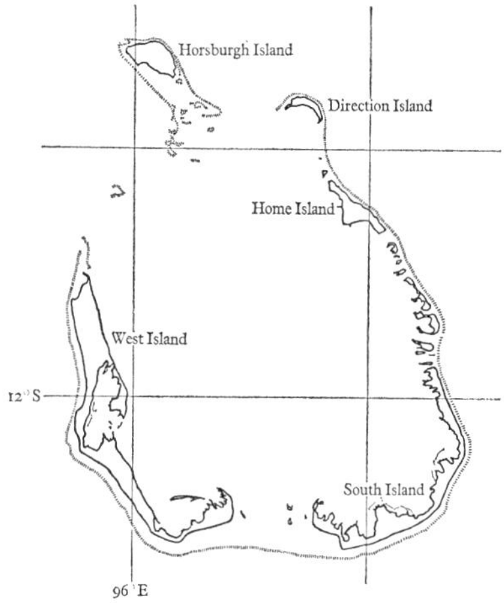

COCOS KEELING

Alexander Hare was not, in any sense, an Emperor. He might, on the other hand, have thought of himself sometimes as a King. What Alexander Hare wanted from his own island refuge was freedom – freedom for sex.

Charles Darwin is the earliest authority on Hare's occupation of the Cocos Keelings. In an interval of making the climactic study of the formation of coral atolls recorded in his journal of the voyage of H.M.S. *Beagle*, Darwin wrote:

> About nine years ago, Mr Hare, a worthless character, brought from the East Indian archipelago a number of Malay slaves, which now, including children, amount to more than a hundred. Shortly after, Captain Ross, who had before visited these islands in his merchant-ship, arrived from England, bringing with him his family and goods for settlement. Along with him came Mr Leisk, who had been a mate in his vessel. The Malay slaves soon ran away from the islet on which Mr Hare was settled, and joined Capt Ross's party. Mr Hare upon this was ultimately obliged to leave the place.

Even epoch-making scientific minds are not always immune from error. Far from being simply a worthless character, Alexander Hare was a splendid and monstrous scoundrel. Ross was his partner. The merchant-ship was a joint venture. The slaves were the members of Hare's harem. Instead of running away, the younger and more attractive among them were most rapaciously seduced by the young and 'unco guid' Scottish artisans whom Ross brought with him, together with his mother in law, to establish his settlement.

In the miniscule setting of what is perhaps the loneliest coral atoll in the world the story of Cocos Keeling repeats something of the story of the earliest days of Rome – with enthusiastic leanings towards the Rape of the Sabine Women.

Of Hare's early years little enough is recorded. In 1805 he was in Portugal as a merchant's clerk. In 1807 he obtained a post with the East India Company and moved to Calcutta. By 1808 he was in Malacca 'living a sumptuously extravagant life'. How he achieved it is not explained, but at this time he met Stamford Raffles, the founder of Singapore, who must have been impressed by him. When Raffles came back to Malacca in 1811, having talked Lord Minto into letting him attempt the capture of Java from the French, he took Hare with him.

The conquest of Java was not a desperately serious campaign; Hare, however, did his share in it sufficiently well – well enough indeed to be sent on immediately to Banjermasin, the Malay Sultanate which, under the Dutch, had exercised control over the greater part of Southern Borneo. His mission, theoretically, was to put down the endemic piracy of the Java Sea; actually it was designed, by Raffles, to secure a quasi-official grip on Borneo.

His campaign against the pirates is one of the minor classics of the great days of Empire building. He was ruthless, fast moving, and ingenious; he trained his crews swiftly and effectively. In an operation that swept luridly across the south coast of Borneo, he wiped out the pirate strongholds one by one and sank the praus that had for generations harassed the native trade and forced even the great East India Company to arm its ships like men-of-war. How much of the loot of the pirates stuck to his own fingers is not altogether clear, but he could have pleaded ancient tradition, one that had existed in Britain at least from the days of Carausius.

It was not, however, his only reward.

The Sultan of Banjermasin, grateful for the liberation of his coasts and harbours from the exactions of an immemorial enemy, decided to reward him personally.

The Sultan, it is evident, was no mean assessor of men. Gold, he must have surmised, Hare had already; jewels, he may have guessed, had been stowed away; but there was, he judged most shrewdly, one other thing that Hare lacked and valued. He made him a present of thirty-four slaves: a tastefully graduated selection of choice young women and girls, selected as much by age as by other qualities, to ensure continuity over a substantial period. There were probably a few men and boys thrown in to do the rough work. Hare, by this time 'Governor and Commissioner for Borneo', was, in all senses, overwhelmed with pleasure.

The possession of slaves was not yet opprobrious. Wilberforce, of course, had been thundering against the moralities involved for a quarter of a century, but the Anti-Slavery Society had not yet been formed and the Emancipation Bill was still twenty years in the future. Hare had maintained a 'household' before Banjermasin, perhaps in Calcutta, certainly in Malacca. It was one of the aspects of 'sumptuous extravagance'. But none of the evidence suggests that he had had more than one or two women at a time. It was the custom.

This was a major harem.

Exactly how much time he had to enjoy it in Borneo is uncertain. His campaign was over well before the area was handed back to the Dutch in 1816, he was already trading on his own account and he had acquired the *Olivia* brig as an armed dispatch boat. His principal lieutenant was John Clunies-Ross, an Orkneyman, mate and chief harpooner of a South Sea whaler, who had arrived at Batavia in 1813 and had there been offered and accepted command of the *Olivia*.

For the last three years John Ross had worked and fought with Hare out of Banjermasin. They even, for a little, defied the returning Dutch. Hare's ambitions were large! So also were those of Ross. With the co-operation of Ross's brother Robert, they built, in partnership, the *Borneo* brig of native teak.

The narrative is here a little interrupted. Hare seems to have moved, with his women, to Batavia, making a dubious personal peace with the Dutch. Eventually he left Java, transporting his harem in the *Borneo*, and went to Sumatra some time in 1818.

Raffles was now Lieutenant-Governor of Bencoolen in Sumatra, already preoccupied with the schemes that crystallized in Singapore. It is possible that Hare had become a little too ripe, even for Raffles. He does not appear to have been offered any part in the great constructive venture of the island. Instead he obtained, with Raffles's approval, manumission papers for his women. Technically they were now free, or perhaps it is more accurate to say that they would have been free if Hare had bothered to show them the papers.

In the *Borneo* he continued to the Cape of Good Hope. He needed the manumissions to get over the difficulties of the Act of 1807 which forbade the landing of slaves anywhere within the British dominions.

Precisely what Lord Charles Somerset, the aristocratic, authoritarian Governor of the Cape, thought of the arrival of his harem is not known. Hare had a presence and, from Borneo, a reputation. As early as 1818 Somerset was having trouble with the powerful colour prejudices of the Dutch. He may have regarded the household with a sardonic amusement. In any event Hare was permitted to stay.

What the Dutch of the Cape thought is not recorded.

For six years he lived in the bosom – or should the word be pluralized? – of his harem. Little is known of this period, but it is plain that public opinion in the Cape turned slowly against him. Probably it was led by one or other of the Churches – the English Church because of evangelical leanings towards anti-slavery, the Dutch because of an established colour prejudice. They are unlikely to have co-operated.

For seven years Hare outfaced it. The details of that private battle must be fascinating but no known record exists. By mid-1825, however, he was at bay. In May of that year the *Borneo* brig arrived at Cape Town, outward-bound to the Spice Islands. John Ross found Hare making preparations to embark on the ship *Hippomenes* commanded by his brother Robert Ross and owned by the partnership.

Shortly after his departure, Hare, with his harem and the miscellaneous children which had occurred, moved to the *Hippomenes*. Hare's battle was, however, not yet over. He was conducting, after nine years, gigantic unfinished campaigns against the East India Company and against the Treasury in London: he had claims for payment for the employment of his ships in Borneo; for compensation for a grant of land at Moeloko in recognition of services; and he had spreading litigation with the Dutch and a vendetta against the Dutch Governor van der Wahl. None of these things had been cleared.

Everything coalesced in a last magnificent feud with Somerset; and Somerset, eighteenth century and autocratic, won. Some time in January the *Hippomenes* sailed.

Ten days later the *Borneo* arrived from the East with a cargo of pepper.

John Ross, coming home with a good cargo, had called, on the way, at Christmas Island. Landing at Christmas Island was difficult. He failed to make it and dropped with the south-east trades down to Cocos Keeling. On 6 December he landed successfully there, carved his initials on a palm bole or two, dug a few wells to test the supply of fresh water, planted a few seeds to see what would happen, and sailed again after four days.

Amongst other things he planted the seeds of certain discontents.

Late in March, Hare, operating in completely independent inspiration, also attempted to land on Christmas Island. Foiled, as Ross had been, by the seas in Flying Fish Cove, he observed the unbelievable cloud of the Christmas pigeons which darkened the sky in the evening and in the morning, and dropped with the south-east trades down to Cocos Keeling.

He found the initials on the palm boles, the wells and the seedlings, and brought his women ashore.

The Cocos Keelings lie 600 miles south-west of Java Head, 1,000 from Australia, farther still from Ceylon. To the south there is nothing except the ice of the Antarctic a quarter of a world away.

Darwin describes the nine-mile lagoon well enough:

> On entering the scene was very curious, and rather pretty, its beauty, however, entirely depended on the brilliance of the surrounding colours. The shallow, clear and still water of the lagoon resting in its greater part on white sand, is, when illumined by a vertical sun, the most vivid green. This brilliant expanse is on all sides divided, either by a line of snow-white breakers from the dark heaving waters of the ocean, or from the blue vault of heaven by the strips of land crowned by the level tops of the cocoa-nut trees.

It is late Georgian but it will serve. The *Hippomenes* dropped anchor and the women went ashore. The goods were landed, the village was established, and the *Hippomenes* sailed. Alexander Hare was left alone on the edge of the green water, inside the 'snow-white' ring of the surf, under the palm fronds that soughed in the trade wind. It was hardly fourteen degrees below the equator but the wind cooled it. The light outside the palms was harsh but under them was shade. The limpid water was made for bathing, the soft noises of the wind for sleep. It had all the undertones of the word languorous, and Hare was farther from other men – save the unconsidered males that he had acquired – than anybody else on earth. There were no rules save those he made, no customs save those he initiated. No moralities except his own.

Emperors have had their flocks of concubines, conquerors have had the pick of cities. But always there were other men about them. Hare had forty women, and all the moon of the tropics, and the knowledge that as the elder women withered the young wheat was ripening.

There was no one to comment, no one to criticize, no one to watch him except the delicate white terns, which hover even today a few feet above one's head on the lonelier islands of the atoll and stare with an incessant and silent curiosity. They must have had much to stare at as Hare tilled, so to speak, his fields.

In this Eden there was no serpent. Even Darwin, ten years later, could find 'only one small lizard, about seven inches long'.

But in the interim peace had altogether fled.

On 27 November 1827 sails were sighted to the westward. Hare's idyll had then endured about twenty months. It had less than three days to run.

In the morning of that day the *Borneo* let go her anchor in 'the most vivid green' water and Hare put out to greet a furiously indignant Ross. The first of what was to be a series of monumental quarrels broke at once. Ostensibly the grounds were that Hare had pre-empted John Ross's rights to the atoll, rights based on the digging of the wells and the planting of the seeds. There was, in fact, room enough for all on the islands. But Ross had brought out ten carefully selected young Scottish artisans, his own six children, his wife and the formidable Mrs Dymoke, his mother-in-law. His purpose was to found a respectable colony – and, very simply, no colony where Hare was could be respectable.

There was one brief attempt at reconciliation. Hare made it. On the morning of St Andrew's Night he sent down rum and roast pig to the young Scotsmen on the island where Ross had settled.

It may have been the rum that was responsible. It may, on the other hand, have been the long and womanless passage. That night the young men, specially selected not only for physique but for the fine

Scottish uprightness of their characters, raided Hare's island – and Hare's women.

Hare sent a note that was, in the circumstances, strangely reasonable:

> I thought when I sent rum and roast pig to your sailors that they would stay away from my flower garden. . . .

It was the end of the beginning. The campaign of the Cocos is one of the more curious small wars of history. Gradually Hare's girls were filched from him. There were moonlight abductions and daring swimming ventures. Canoes slipped in. Boats carried out blatant piracies.

Slowly Hare was driven from his island to another, and to a more distant island after that. There were spells of peace and intervals of amity, but always battle began again. In the end he shut himself in a timber fortress on the sand knoll of Prison Island with what remained to him.

Some of the abductions were real, some were contrived. Hare was growing older now, the young girls looked to youth. There are even today many legends in the islands of this time; there was much room for legend. The date of the ending of the campaign is unrecorded but it dragged on for more than seven years.

In the end Hare left.

He had lost his island, his village, his women, his children; there was nothing now that was his on Cocos Keeling. Uncertainty even touches his own ending. But out of the obscurity of it there comes at last one sentence like a sword. I found it enshrined for all time in a document in the old India Office Library. It says:

> Alexander Hare died in an attempt to establish a new harem.

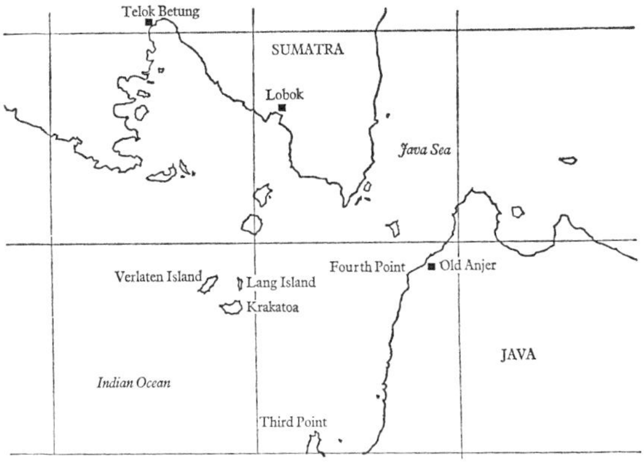

Terror

KRAKATOA

The Thousand Islands lift first as you approach Sunda Strait from the north-east. There are, in cold fact, only eighty of them, but it scarcely matters – eighty are enough to cover the horizon. After them you see Java and the coastline about the Bay of Bantam, and it grows. It is very green, very rich, inordinately ancient. Then you begin to pick up Sumatra, and it is blue at first and high and distant. Afterwards the outline solidifies, and narrows until the water gap is barely fourteen miles wide and you pick up Dwars-in-den-Weg, the island in the centre – Slap-in-the-Middle would be an adequate translation – and Sunda Strait divides into two four-mile channels, and you slip through over the smooth blue water, and the Strait opens again.

Krakatoa lies there.

The name comes from Rakati, the antique Malayan word for a crab. Before the disaster the group was made up of the main island, Krakatoa, Verlaten Island to the north-west, and to the north-east Lang Island – the claws, as it were, of the crab. There was in addition a small islet called, improbably, Polish Hat.

Krakatoa had an area of about eighteen square miles. At its southern end rose the cone of Rakata, the main crater, 2,677 feet in height. There were two lesser cones – Danan in the centre of the island and roughly 1,500 feet, Perboewatan, insignificant at 399 feet – and a number of smaller fumaroles.

To the north-east were two other islands, Sebesi and Sebuko, separated from Krakatoa by the Besi Channel, one of the two main fairways through the Strait.

Sunda is the gateway of the Far East. Ships use it to avoid the long and difficult passage of the shallows of the Straits of Malacca or the awkward dog leg of the haul to the east for Bali Strait.

The islands were very beautiful, covered with high forest. Seamen coming in from the slow voyage across the Indian Ocean with its unalterable blues and its unceasing glare found them 'a wonderful relief to the eyes'.

In May 1680 Krakatoa erupted.

According to the Dutch, who had just secured the principality of Jakarta which borders the Strait to the east, earth tremors early in May increased in severity and in frequency until, towards the end of that month, Krakatoa began to put up enormous quantities of pumice, which in due course 'covered the sea'. Dust, heat, and the gases of the eruption destroyed all vegetation.

It was not regarded as important. Krakatoa was uninhabited. Only the villagers of the coasts on either side of the Strait bothered to visit it for what they could collect of tropical fruits or for the fishing off its faintly inhospitable shores.

In any event, there were other and more active volcanoes in the East Indies. A great fault runs with the main mountain chain of Java. It enters Sunda close to Third Point, alters course slightly to the north at Krakatoa, and ultimately underruns the main chain of Sumatra. A cross fissure runs from just behind Java Head through Princes Island to an intersection precisely below the northern half of Krakatoa. Major volcanoes lie along these lines: Pajang above Java Head; the peak of Princes Island; the three cones on Krakatoa; Sebesi, which rises to 2,825; and Raja Bassa on Sumatra, tallest of all, 4,398 feet.

The Dutch, in 1680, were unaware of the threat explicit in the fissure, nor, for the next 200 years, was there cause for alarm. The islands grew green again, the fishermen returned. The Strait was peaceful.

In 1877 earth tremors offered a new and barely perceptible warning. As in 1680, they increased gradually in frequency, and in the September of 1880 the lighthouse at First Point – on Java Head – was destroyed.

Krakatoa itself gave no indications of activity.

Three years later in the early morning of 20 May 1883 'sounds like the firing of artillery' were heard a hundred miles away in Batavia and Buitenzorg. Their source was not at first identifiable but it was apparent that they came from the Sunda area. On 21 May fine ash fell at Telok

Betong, forty miles to the north, and at Buitenzorg, a hundred miles to the east. In the early evening a giant column of steam surged up out of Krakatoa. 'Fiery flashes, loud explosions, and a vast discharge of pumice and dust, with vivid flashes of lightning' completed the picture. By 24 May dust was recorded at Timor, 1,300 miles down the chain of islands to the eastward, but after this the force of the eruption diminished again. 'People stopped paying attention.'

Not all the people.

On 26 May 'an excursion party' was got together at Batavia 'and proceeded in a steam vessel to the scene of the eruption'. The eighties were the heyday of the intrepid amateur. The party reached Krakatoa on the morning of the following day and discovered that the peak was no longer green. Over the entire north of the island the trees were defoliated, and a white dust covered everything 'like snow'. Rakata itself, however, was not in eruption; Perboewatan, the insignificant 400-foot crater at the northern end, was manifestly the danger point.

With characteristic recklessness the party landed through dust and broken pumice, climbed the treacherous surface to the crater lip and discovered that it had expanded to an enormous mouth 3,000 feet in diameter and that, from a vent about 150 feet across in the crater bottom, 'steam appeared with a terrific sound'.

They had taken a primitive camera with them. The colour plate published in the Royal Society's masterly report on the eruption is drawn from one of the photographs brought back. It shows the glare above the crater, the steam and the smoke, and the stripped tree boles of the lower slopes.

For a brief period activity diminished. Then on 24 June a second column of steam appeared, and when on 11 August the Dutch topographer Ferzenaar visited the island, he reported forests totally destroyed, twenty inches of dust everywhere, and three columns of steam – one of them now rising from Danan, the centre crater. Eleven smaller fumaroles and vents were in operation.

Whether Ferzenaar entirely appreciated the situation is doubtful. Geologically the Krakatoa group was what Darwin called 'the basal ruin' of a vast ancient volcano, twenty-five miles in circumference. Krakatoa itself was the plug which stopped its throat.

The ridges of the three islands were – and are – the surviving fragments of that crater lip. The low ground of the northern two-thirds of

Krakatoa and Polish Hat were part of the sequential eruption of the main cataclysm, which may have split Java from Sumatra as the sea flooded across the depression that followed a vast explosion.

What Ferzenaar did determine was that one by one the lesser craters of the plug were flaring into activity. His was the last official visit.

Up to that time Krakatoa was in what was then called the Stromboli phase – a period of moderate activity. It was about to pass into the Vesuvian phase, the period of paroxysm.

On the night of 24 August Captain Watson, master of the British ship *Charles Bal*, on passage to Hong Kong, sighted 'a strange white haze or silvery glare' on the bearing of Sunda Strait. On the following night it was again visible. At dawn on Saturday, the 26th, he sighted Java Head, the wind served him for the Strait, and at nine o'clock he cleared Princes Island and made out the enormous column of smoke and steam above Krakatoa.

Watson carried on – he could not afford to lose the wind. At noon he was south-west of the smoke and going well. At one o'clock there was a fantastic detonation – the first of the great sounds of Krakatoa. It was heard in Buitenzorg, a hundred miles inland to the eastward, it was heard far up Sumatra, but the *Charles Bal* maintained her course. In mid-afternoon, with Krakatoa due north, distant ten miles, Watson lost sight of the Java coast. At Buitenzorg it was dark already, Batavia was closing up. Pumice was falling with the ash. Occasionally through the murk, the lightning and the flashes of Krakatoa were visible but navigation had become impossible. Captain Watson shortened sail and began a long series of tacks, trying to hold a safe course between the half-seen volcano and the unseen land.

Out to seaward in the clear air the ship *Medea* measured the summit of the smoke cloud as a height of seventeen miles.

All through the terrible night the detonations went on. All through the night the dust fell, and the pumice coalesced in floating islands on the surface. The *Charles Bal* survived. At 6 a.m. on the morning of the 27th Watson sighted the coast of Java. His log entry reads: 'Continued on voyage.'

The steamship *Gouverneur General Loudon* left Batavia on the morning of the 26th. At 2 p.m., shocked by the one o'clock detonation, she put into Anjer on the Java coast inside the Strait, picked up passengers and headed north-west between Sebuko and Raja Bassa for

Telok Betong, at the head of Lampong Bay. Her course was thirty miles clear of Krakatoa, but pumice and mud covered her decks and shone luminous in the darkness. The crew fought feverishly to clear 'the work of evil spirits'.

At 7.30 p.m. she was at Telok Betong. At midnight conditions were dangerous, communication with the shore impossible, and Commandant Lindeman steamed out into Lampong Bay and anchored, advising the naval steamer *Berouw*, anchored close to the pier-head, to do the same.

At 7.30 in the morning, with a gale of wind blowing, *Loudon* headed out into the rain of pumice; by ten o'clock she was stopped in heavy seas, but later resumed course.

The gunboat *Berouw* decided not to move and lay enduring the filth and the growing discomfort of the seas at anchor in the port. At about half past six, in what should have been first light, a giant wave underran her, but her anchors held. At about a quarter to eight another great wave lifted her – the anchors still held. At a little after half past ten a vast sea climbed out of the murk and the *Berouw* was lifted clear of her anchors, swept with inconceivable violence almost two miles up the valley beyond the town, and left, wrecked, thirty feet above sea level.

She did no damage to the town in passing – there *was* no town now!

The *Loudon*, heading well clear to the west of Krakatoa, made her way back to Anjer. Grimly her log entry reads:

> Called at the part of the coast where Anjer formerly stood and then proceeded to Batavia.

There were, on this terrible Sunday morning, four cataclysmic explosions: in the dawn at half past five, at a quarter to seven, at ten o'clock, and at ten fifty-two.

The explosion at ten o'clock was the loudest sound ever heard by man. It was heard across all Java. It was heard in Borneo, where at St Lucia 'the natives inland murdered poor Witti when they heard the sound', fearing that they were about to be attacked. It was heard at Singapore, where two steamers were sent to sea to look for a vessel in distress. It was heard at Achin in Northern Sumatra, a thousand miles away, where it was thought that the fort was being attacked and troops were put under arms.

It was heard, inconceivably, at Rodriguez Island two-thirds of the way across the Indian Ocean, 3,000 miles from Sunda Strait.

It was the sound of the plug in the throat of the old volcano blowing out. Two-thirds of Krakatoa went with it, an estimated 200,000,000,000 cubic feet of soil and rock and pumice and lava. The tidal waves that followed upon the sound travelled farther still. Evidence of the movement of the titanic wave that followed the ten o'clock explosion was detected near Cape Horn, traces were claimed at Cherbourg in the English Channel.

The shock wave of the explosion spread in a vast circle that enveloped the world, travelled to the antipodean point of Krakatoa near the Gulf of Mexico, and doubled back on itself again and again.

The pumice, so deeply piled at first that ships could scarcely make their way through it, fetched up in the months to come on the shores of every ocean in the world. And in the Sunda Strait the innumerable small craft everywhere were driven with their people far ashore, the wreckage of the coastal villages was piled on the slopes of the hills, the ports were wiped out. At Anjer the first wave was thirty-three feet in height. No one was left to record the second. At Princes Island the great wave was fifty feet, at Telok Betong seventy-two, at Kalembang eighty, at Merak where it passed the narrows it was estimated by two men – Nieuwenhuis, who put it at 100 feet, and McColl, who judged it to be 135.

Thirty-six thousand three hundred and eighty people died in it.

There was one other phenomenon, innocent and astonishingly beautiful.

In June 1883 'fine red sunsets' were reported in the Alps. In July they were being noted at widely separated points. By November they were everywhere – Australia and Iceland, North America and India, all over the Pacific, all over the Atlantic. A series of rapid colour sketches made across the river at Chelsea in London on 26 November at intervals of a few minutes and published in the Royal Society's report shows the all but unimaginable speed and complexity of the colour changes. At 4.10 p.m. it was a soft red with areas of yellow, ten minutes later it was almost wholly red, at 4.30 it was predominantly yellow, at 4.40 orange, at five o'clock red with great angular flares of scarlet, at 5.15 a violent and unbearable crimson.

Week after week through the winter the dawns and sunsets were spectacles of inconceivable fantasy. Not until April and the spring did the brilliance fade.

Then, with the slow, imperceptible falling of the spicules of volcanic glass, the threads of intangible crystals, the minutely powdered pumice that Krakatoa had blasted into the troposphere, the horror of the Sunda Strait came to an end in beauty.

ITSUKU-SHIMA

Natural forces engendered the horror of Krakatoa. Horror may also be engendered by man.

The Inland Sea of Japan has in still weather a balanced perfection of form and colour. There are days in its summer when the sun is diminished in a quiet haze. The water and the fishing boats and the sculptured islands are nacreous in the softness of that light. As they fly silently towards a moving ship, they seem to change their colours as the colour of a pearl shell changes in the movement of a hand.

It is at least to be believed that it was in such a light that the Goddess Ichikishima-hime and her companions made the founding circuit of Itsuku-shima.

Itsuku-shima lies close against the high shore of Honshu in a northward-reaching gulf of the Inland Sea. The Strait of Onoseto which separates it from the mainland is calm and narrow. A thousand-year tradition declares that the Goddess crossed it by boat, coasted to the north, rounded the little cape of Hijirizaki, and landed for the first time at the ancient shrine of Suginoura, landed again at the crescent beach of Takanosuura, rejected each in turn as a possible resting place and sailed on round the wooded coastline of the island, calling in all at seven places before she came to the pool of Kagami-no-ike almost at her starting point. On the pool of Kagami a temple was built.

After twelve centuries the floating temple of Itsuku-shima (it is called also Miya-jima) endures as one of the exquisite achievements of the hands of man. It does not, in fact, float. It offers, instead, the perfect illusion of floating. Piles in the still water hold it, as piles supported

The water colours of the Krakatoa sunsets painted for the Royal Society from the Chelsea embankment are amongst the most remarkable records of the cataclysm that destroyed the volcanic island in the Java Strait.

Towers of Manhattan – a helicopter's eye view of the fantastic cityscape of New York.

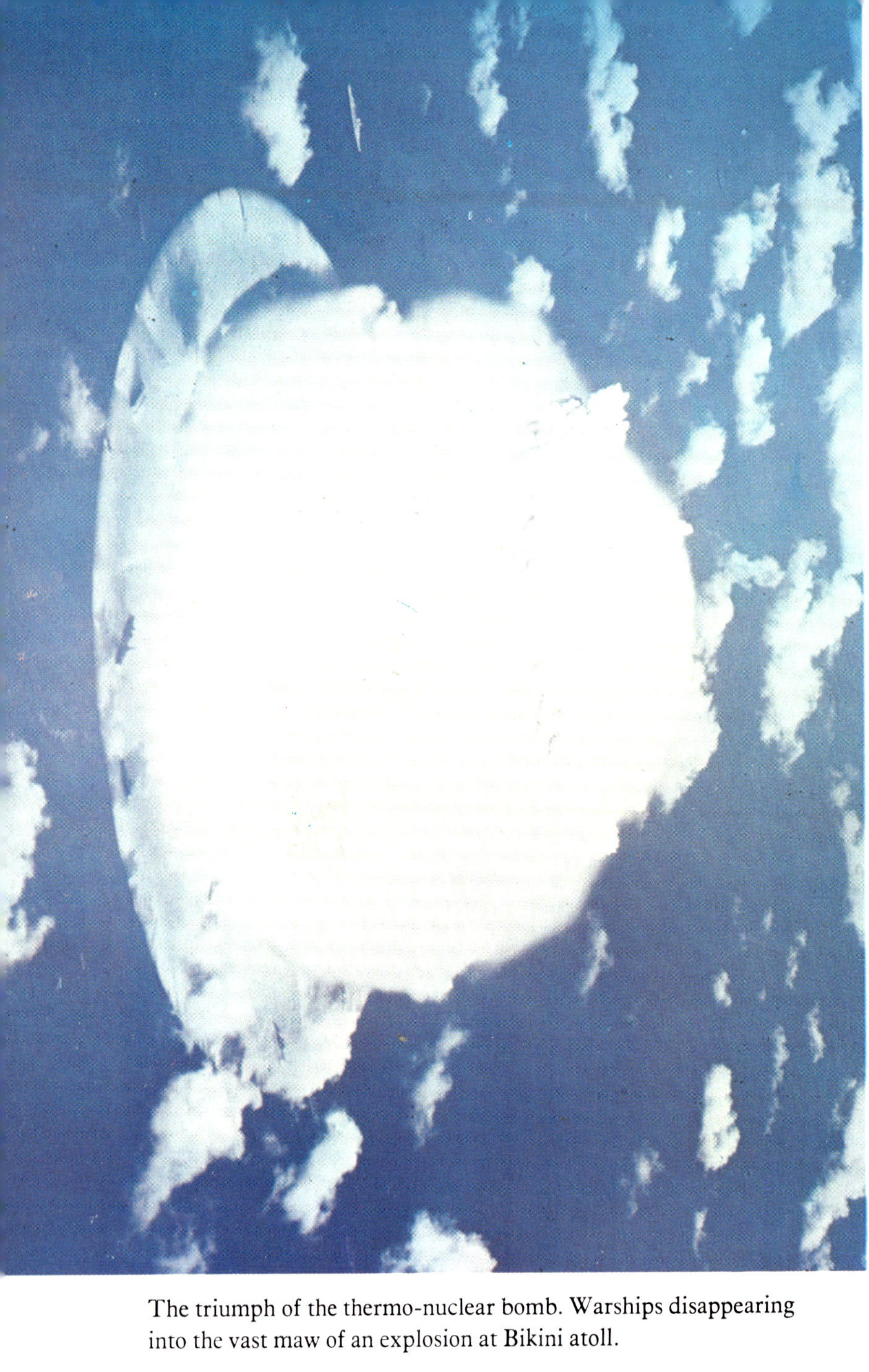

The triumph of the thermo-nuclear bomb. Warships disappearing into the vast maw of an explosion at Bikini atoll.

the lake villages of Neuchatel, of Constance and Bienne in prehistoric times or on the Amazon and in Borneo and Central Africa today.

It has not the permanent solidity of the churches of Venice – it is built wholly of wood – nor does it carry the arrogant defiance of Notre Dame – jutting against the current of the Seine – but it is older than they are, none the less, and may continue so. It does not aspire to the triumphs of the vertex. It was conceived very simply in a perfect appreciation of the horizontal. A delicate and cerebral complex of lines parallel with the level water, level with the parallel shore, underscoring the mass of the tree-laden peak of Misen. Level galleries, superbly proportioned, link its pavilions, level roof ridges crown its halls, only the hollow curve of its roof tiles supplies the contrast in a geometrical balance that has no precise counterpart anywhere else in Japan.

It is a quiet temple in a quiet island. In a sense it was dedicated to silence. As in Delos, the ritual of birth was not permitted here nor, in so far as men could assure it, the ritual of death. No one was buried here, and even the mourners who went to the mainland for the ceremonials were not permitted to return for fifty days. There were no vehicles here, not even the jinrikisha. And there were no dogs.

The birds sang, the crickets shrilled, the gongs and the bells of the temples – Buddhist as well as the great Shinto temple on the water – sounded. There were also the drums and the music of the Nō dance, which was here preserved through the centuries of its eclipse in the wider world. With its masks and its vestments and its immemorial ritual movements like the wing beats of a tired swan, it had come from Chosen, or India, or China, in the seventh century, and on this island it remained alive. These were the sounds of Itsuku-shima – these and the boat music when the pilgrims made the traditional circuit of the island, the priests and the musicians in the leading boat, the pilgrims in the centre, and the cooks and the food in the third.

The temple lives with the rhythm of the sea on which they sailed. Twice a day the tides change under it, altering its proportions, deepening its façade, changing the colour in which its rests. Sometimes in the low water of spring tides it stands high, borne aloft on its columnar piles. Sometimes it is at rest, inseparable from the surface of the sea itself.

In the bright August of 1945 it was more silent, perhaps than it had been at any time. There were rare pilgrims, but the great festival of

17 June had brought few to the island. The railway services were cut to the limits of necessity, endlessly disrupted by bombing raids from the American aircraft carriers. Petrol was unobtainable, substitutes were inordinately costly. Ship traffic on the Inland Sea had been cut to a minimum by mining from the air. Even the *ban-sen* – the small junks – which had been given a new lease of existence by the failure of modern transport, were growing scarce. The great cities which sent the major part of the pilgrims to Itsuku-shima were now the target of the war. Kobe was shattered, Osaka a ruin, Tokyo itself a desolate ring of modern buildings round the Imperial Palace in a waste of burned wood and shattered tiles.

Itsuku-shima was undamaged. From across the water at Kure it heard endlessly the crash of anti-aircraft guns and the deep reverberation of bombs as attacks on the great naval base came in from the outer seas. Down to the southward it heard the strikes on Iwakuni and its airfield. American aircraft had become a part of the scenario of defeat, at times it seemed that the sky was never empty of them – the sharp roar of the Navy's Avengers in the daylight, the deep drone of the B-24s and the B-29s at night and in the early morning.

It had become worse since Iwo Jima fell.

On this August morning there was a handful of people on the summit of Misen – priests, workmen, and two or three soldiers from the look-out post. There were more people working on the high slopes above Suginoura in the north and above Takanosuura, the bathing beach, but not many, according to the local people.

Early in the morning there had been an air raid warning on the mainland. A single American B-29 bomber, flying high, had raced past the head of the gulf, continued westward for ten miles, turned and flown back. Afterwards it disappeared. It was not important; American reconnaissance aircraft were a fact of life. At 7.31 the all clear sounded.

At 8.06 the radar station at Matsunaga detected two more aircraft and, moments later, a third. They were reported to Air Defence. No action was taken.

On Itsuku-shima they did not see the aircraft – they were invisible to the north-east when Matsunaga picked them up. Nobody would have taken notice in any event. They heard them later, however – the carrying engine-note of two B-29s, at altitude.

Looking towards the sound at least two people claimed to have seen specks in the sky. Not aircraft but specks that drifted. They did not identify these as parachutes.

Those in the open saw the flash – all of them, even those who were turned away from it. 'It took', one man said, 'the light from the sun.' Some saw colour in it, blues and in one instance green. This may have been subjective. Most of them saw only brightness, a monstrous intolerable and unbelievable brilliance – the light of the first atomic bomb, shattering the sky over Hiroshima ten miles to the north.

Whether anyone on Itsuku-shima saw the second aircraft sweep round to the left and head for a brief moment close to the open water of Hiroshima Wan is uncertain. This was *Great Artiste*, flown by Major Charles Sweeney, who had dropped the recorded instruments and transmitters with their parachutes a little to the left and astern of the bomb. *Enola Gay*, piloted by Colonel Paul Tibbets who dropped the bomb, had turned through the north and headed eastwards until she was clear.

They had precisely forty-three seconds in which to escape the lethal field of the thing they dropped.

Hiroshima had precisely forty-three seconds in which to live.

It was not an ancient city – not as old as Itsuku-shima, for example, but old enough. Mōri Terumoto, the warrior, founded it when he built the castle in 1594. It was the stronghold of the Asano family through the violent history of the Shogunate. With the restoration of the Meiji, the castle decayed, only the Tenshu-kaku – the five-storied keep – and half its inner moat remained. But in the war of 1894 it was established in power again as Japanese military headquarters, the Emperor took up residence there and a building for the Diet was erected in the grounds. From Ujina, its port on Hiroshima Wan, the armies sailed for China.

In 1900, and again in the Russo-Japanese War of 1905, it was the major base for military expeditions. In 1941 the Ever-Victorious Fifth Army sailed from it to the capture of Singapore, and when with the defeat in the Pacific and the drastic warning of the loss of Iwo Jima the Japanese Home Defence Command was reorganized, Hiroshima was designated Second General Army Headquarters to control the south-western half of Japan. Headquarters were again established in the Mōri castle, and preparations were made for the Emperor to withdraw to it when Tokyo was threatened. It was not a city of peace, ever, but a potent city of war.

Strangely, in the crescendo of the general air assault on Japan, it was ignored. Two small bombs were dropped by mistake in a naval strike in March 1945. In May a lone B-29, which had missed its planned target, loosed ten 500-pound bombs across it without serious effect.

It was not, however, set aside as an atomic target until 3 July, when the Joint Chiefs of Staff ordered that four cities for the present should not be bombed: Kyoto, the ancient capital with its treasure of palaces and temples; Hiroshima, with the South-Western Headquarters; Kokura, the great military arsenal; and Niigata.

Kyoto was dropped from the list at Secretary of State Stimson's intervention on the grounds that its destruction would have been the ultimate vandalism; Niigata was omitted as too small and unimportant, Nagasaki was inserted as an alternative, and the list was approved by the Potsdam Conference.

Hiroshima was declared the primary target, but the final selection was to be dictated by cloud conditions over the three cities as Tibbets and the *Enola Gay* approached the coastline of Shikoku and the Inland Sea. The formal signal 'Advice: Bomb Primary' from the leader of the reconnaissance force determined its fate.

Hiroshima was covered by anti-aircraft defences. It had an elaborate shelter complex. Since the spring it had been developing a vast fire break system which had involved the destruction of thousands of houses. Inadequate administration and a degree of individual obstruction, despite the brutal warning of the Tokyo fire raids, had delayed its efforts, but in the event none of these things was important. All of them could not have saved Hiroshima. The single bomb that dropped from a single plane killed 78,500 men, women and children.

On Itsuku-shima those who could still see clearly after the flash watched in horror as the mushroom cloud built up on a vast stem of dust and smoke. As the flames leapt in it, the sound reached them. The loudest sound since Krakatoa. The most appalling sound ever made by man. It crashed over the quiet island – the sound of a civilization changing.

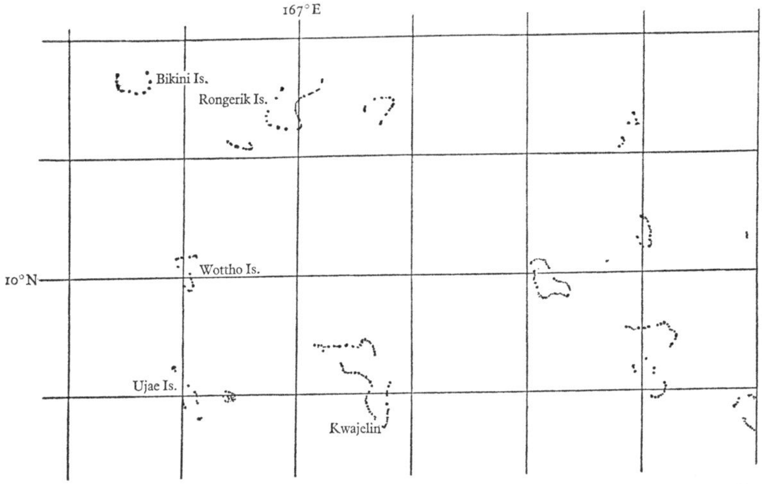

BIKINI

The Marshall Islands lie in a vast ellipse that sprawls across 400,000 square miles of the Pacific. They have a complex pattern. A score of atolls, most of them of enormous size, form two irregular chains that run from north-west to south-east. Lukunos, at the southern tip, is 750 miles from Eniwetok. Outside the chains are ten other atolls and a dozen separate islets. They are as scattered as a constellation in the night sky. In all the immensity of the group there are just 160 square miles of solid ground.

When they were first found in the course of the great canoe migrations is unknown. The Spaniard de Saavedra was there in 1529, and the English Captain Wallis 250 years later. In 1885, in the years of the land grab, the Germans seized them and in 1914 used them strategically, once. Von Spee sailed from them after refuelling on his way to victory at Coronel – and death off the Falklands. After World War I Japan was given them, under mandate.

They were doubtfully profitable. In 1941, however, the submarines of the Pearl Harbour Attack Unit fuelled at Kwajalein, and the group became important as a defensive area in the line that began at the Kuriles and swept south through the Carolines to Timor and New Guinea.

In February 1944 Admiral Nimitz struck suddenly at the Marshalls in one of his great thrusts across the Pacific. Majuro was taken unopposed, Roi and Namur fell after small but bloody battles, Kwajalein was stormed, and at the far end of the chain Eniwetok, with its three defended islets, was overrun. The Japanese lost 11,000 men in a fruitless defence.

In the islands where there were no Germans, no Japanese and no Americans, life was elegiac.

All atoll islands are the same – and all are different. Darwin's description of Cocos Keeling will serve for the Marshalls; one island will have a narrower beach, one atoll a deeper lagoon. The reef will be farther out with some and the sound of the surf less insistent. One will have only an exiguous screen of palms and another will be deep enough to winnow out the salt from the trade wind and allow deciduous fruits to grow. All have one thing in common – the enchanting beauty of the light through the boles of the outer palms. It is as exquisite as the morning light through the stained glass of a lancet window in a great cathedral. Simpler in its pattern, it is at least as brilliant in its colour. High up it is the deep cerulean of the ocean sky, flecked often with the trade wind clouds, framed in the drooping curve of the palm fronds. The blue is held in the mullions of the boles while it diminishes, until near the horizon it is almost white. At the horizon it is answered by the purple blue of the deep sea, and below the line of the sea horizon is the parallel horizon of the surf, blinding white and violent in its motion. Inside the surf is the pale green of the shallows of the reef, and inside that the lesser white of the beach. It is a stained-glass window composed in horizontal bands, and it is always the same.

The people of the Marshalls loved their islands and they loved the calm of their lagoons. There were not many of them. In 1872, when the land grab was getting under way, Majuro and Arlino each had 3,000 people, but big atolls like Otdia and Kwajalein had barely 200 each and Namur only 150. Eniwetok had forty.

When, in March 1946, the people of Bikini were persuaded to leave their atoll, there were 167 of them.

They were a gentle people. Bikini was as quiet as Itsuku-shima – not as sophisticated, but as quiet. Its Chief, Juda, after he had discussed the matter with his people, said mildly: 'If the U.S. Government needs to use our houses for the goodness of mankind, then by the kindness of God we are willing to go.'

So a naval vessel picked them up, with their mats and their cooking utensils and their canoes, and whisked them overnight to Rongerik, 150 miles away, where they were desperately unhappy.

The vital question is the definition of goodness.

The development of nuclear weapons, from the uranium bomb that obliterated Hiroshima to the multi-megaton thermo-nuclear weapons of today, exists as a case history for the study of the process by which science and technology are taking over man. Most of those who took part in the higher levels of the Manhattan Project and its successors have written or discussed their shares in terms ranging from an absolute conviction of right to accents of unshrived guilt. Few of them would concede that they were in any sense prisoners of either a scientific equation or of a machine. Yet there is in the history of thermo-nuclear success an inevitability, at all stages of development, that has a logic which is altogether inhuman.

After Rutherford's Cambridge experiments, the work of his successors is in a sense automatic – it was quite certainly inevitable. The climactic splitting of the atom by Hahn and Strassmann in 1939 was effectively the end of a beginning, clarifying what had gone before and illuminating what clearly had to come. It was the perception of its future threat by the group who were sometimes known as the immigrant scientists – Szilard, Wigner and Teller in America – that persuaded Einstein to write his germinal letter to President Roosevelt. It was the fear of a German atomic bomb that led directly to the inception of the Manhattan Project. That the German bomb failed to materialize, in no way alters the contention. Because Hahn and Strassmann's work indicated that a bomb was theoretically possible – the bomb had to be.

The decision that was made almost simultaneously with Pearl Harbour to begin all-out research and development was not directed specifically at the Japanese. The results were the same. On 2 December, when Enrico Fermi and his team produced the first self-sustaining nuclear chain reaction in the squash court of the University of Chicago, the Atom Age began.

History in that Age was, and continues, irregular. I was on the island of Guam when the Hiroshima bomb was dropped, two and a half years later. Three days after, as I left the mess hall, I saw one of the Manhattan Project Commission people on the island and asked if it was reasonable to go for a bathe. He said: 'Stick around!' I asked point-blank if it was another 'Hiroshima' bomb. He said: 'Hell no! That's obsolete now.'

The switch from the Uranium to the Plutonium bomb was made at Nagasaki. Man was advancing.

After Nagasaki there were no more targets.

Nuclear science, however, had acquired an imperative. There were new variants to test. A proving ground was clearly an essential.

Eniwetok atoll was designated. It was distant from other islands in the Marshalls, immensely far from continental land. It had a vast lagoon, and its population was minute.

Before the Hiroshima bomb was dropped, long before the end of the war, Edward Teller had become totally involved in theoretical speculation as to the possibilities of thermo-nuclear fusion. The precise balance between the contribution of his individual genius and the pressures of the nuclear age in which he had his being may not ever perhaps be possible to determine. It seems certain that had it not been for the fact that he was immersed both in the science of nuclear physics and the military technologies that stemmed from it, he would not, in his own lifetime, have reached this point. Nuclear weapons development had, under the Manhattan Project, acquired a dynamic of its own.

By the spring of 1951 his energy and his enthusiasm had induced authority to set up an experiment to test the principle of his theories. Called Operation 'Greenhouse', it was carried out on Eniwetok atoll, and it was experimentally successful. In June he was able to prove to an initially sceptical audience at Princeton that a thermo-nuclear bomb was feasible.

By 1952 Operation 'Mike' had been set up on the island of Elugelab to test the practicality of a bomb.

The test rig was large and elaborate, it required complex refrigeration because the stage of miniaturization was still hardly begun, and the device was not dropped but housed in a building on Elugelab.

On the morning of 1 November 1952 watchers, forty miles away from the island, saw a fireball form, three miles in diameter, in a fraction of a second. Man had not ever made a brighter light.

When the steam and the smoke of the mushroom had disappeared, a photographic plane flew over to look at Elugelab. The photographs it took showed a vast hole in the reef and the floor of the lagoon a mile across. The water in it was as dark as the blue of the deep sea beyond the reef.

Elugelab itself had dematerialized.

A third test was scheduled.

It was now no longer possible to use Eniwetok. There were, of course, other islands than Elugelab, but they lay in the eye of the winds, west and a little south of the northernmost islands of the chains, and the problem of fall-out had suddenly acquired a singular significance.

It was at this point that the people of Bikini were asked to leave.

In size and weight this device met the requirements of a bomb. It also was not dropped, however, but exploded on a tower. It was estimated that it would produce a force of eleven megatons.

It was altogether successful – except that it produced a force of fifteen megatons, and that a freak of the trade wind carried off the fine dust and the fragmented coral of the explosion, which was now radioactive. They drifted across an American destroyer and the islands of Uterik, Rongerik and Rongelaap (east and a little south of Bikini atoll) and, passing on, dropped, some of them, on the deck of the Japanese trawler *Fukuryu Maru* – the Lucky Dragon.

Winston Churchill said of it:

> . . . it might well be that, by a process of sublime irony, we shall have reached the stage in this story where safety will be the sturdy child of terror, and survival the twin brother of annihilation.

Between 1946 and 1958 there were twenty-three nuclear explosions in the Bikini atoll. In the 'Bravo', the biggest of them all, two islands were dematerialized and a third cut to a sand bar.

After 1958 this atoll was abandoned: it was radioactive.

The people of Bikini asked to go home. Rongerik, their first stopping place, had been altogether a failure; it was smaller than Bikini, less fertile, and its lagoon was unquestionably inhabited by a female demon.

They were moved to Kili, a speck of land between Jalut and Namorik, with no lagoon at all. They remained unhappy. Some of them spread out among the other islands of the group; their numbers increased; there were gifts from America and, in due course, pensions. But they still pined for Bikini – they used the word *Lamoren*. The word means ancestral land; it has an ineradicable nostalgia.

In August 1968 President Johnson announced that Bikini atoll was at last clear of radiation. They could go back.

Nine men went first, for a formal repossession.

Over much of the islands no palm trees grew. The beaches were covered with debris, parts of aircraft, concrete block-houses, the ruins of the chamber in which the bombs were assembled. There were warning signs at scores of points. Broken living accommodation.

The leader of the men asked gently: 'What have you done to our atoll?'

It is a commentary on the levity of this technological era that the name of that most tragic island was at once transferred to a frivolous article of feminine beach-wear.

Trade

SINGAPORE

The singular advantages of islands in the development of trade were seized on by the Phoenicians in the earliest years of their succession to the control of the Mediterranean. Whether the eruption of Santorin devastated the maritime economy of Crete or whether other factors contributed is immaterial. What is beyond question is that in the eastern basin of that sea Mycenae succeeded Minos, the Dorians destroyed Mycenae, and thalassocracy passed in due process to Phoenecia.

From their first arrival on the shore of the Mediterranean the Phoenicians had looked to the sea. Sidon was a promontory fort, protected by water on three sides and by defensive works across a narrow isthmus. Tyre represents a logical advance by a people in possession of effective boats – an offshore island close enough to the shore to make boat connection possible, distant enough to provide sheltered water for a harbour, secure enough to provide protection against an enemy without sea skills. Between them the two supplied the stock pattern for future expansion. Carthage was a peninsula that was almost an island, Gades – Cadiz – was an island that was almost a peninsula.

Three thousand years later and 4,000 miles away, Thomas Stamford Raffles employed the same basic principle in determining the establishment of Singapore. There had been, it is recorded, a walled city on this offshore island centuries before Raffles landed there, but Singapura – the City of the Lion – had disappeared when the empire of Sri Vijaya foundered at the end of the fourteenth century. In 1819 there was a fishing village only.

Singapore island is remarkably like the Isle of Wight in size and shape – an irregular diamond a little longer than the Wight, almost exactly as wide, and separated from the mainland by a dog-legged strait narrower than the Solent and Spithead but for its purpose wide enough. Singapore city lies approximately where Ventnor lies. It lacks the cliffs of Ventnor but it has a natural harbour.

Emperors have commanded the establishment of cities; dictators, using the resources of nations, have decreed them. The vast entrepôt of Singapore was created by a junior official of a trading company who had not seen its site before he landed there, who visited it three times only, and whose longest stay lasted a bare eight months.

Raffles was the son of an incompetent merchant in the West India trade – slaves out, sugar and molasses and tobacco home. At the very height of the West Indies boom his father contrived to fail. Raffles, at school at Hammersmith, was placed at the age of fourteen as a clerk with the East India Company on a salary of £10 a year. When after two years his father died, that salary had to support himself, his mother and a horde of sisters.

Ten years later he was still at approximately the same pay level, but in the interim he had educated himself, he had acquired a singular fund of knowledge of the East, and he had also been accepted in the circle of William Ramsay, head of the East India Office. When, in 1805, the Company determined to strengthen its lines of communication with China by establishing a port of call – and a new Presidency – at Penang, in the northern entrance to the Straits of Malacca, he was appointed assistant to the Chief Secretary, with an astonishing advance in salary to £1,500 a year.

The selection of Raffles for the post was appropriate; the choice of Penang was inept – it was not a focal point, and it did not tap automatically a hinterland with a significant trade. It was, however, the only potential deep-water harbour in the possession of the Company in the area, and the politics of the East Indies were in a state of flux. Napoleon's seizure of Holland had transferred the Dutch possessions in the Far East to France. Java in the hands of the French was capable of closing the Sunda Strait, the direct gateway to the Far East on the route via the Cape of Good Hope, and of threatening the movement of traffic from India through the Straits of Malacca to the China Sea.

Malacca itself, the only visible alternative to Penang, silting up and decrepit, had accrued to the Company in the carve up of Holland's eastern empire, but to the commercially minded Court of the Company in London it was altogether too likely to be handed back to Holland in the terms of an eventual peace. Its position, at the narrows of the Straits, was in any event unsatisfactory.

Raffles sailed for Penang in 1805, married, on the strength of his rise in salary, to Olivia Fancourt, one-time mistress of Thomas Moore, the poet. After eighteen months he was promoted Chief Secretary, by 1810 he was an accepted authority on the Indonesian area and had established, through the Orientalist John Leyden, a personal contact with Lord Minto, the new Governor-General in India. In that year, bursting with a plan for securing the East Indies for the Company, he made a hideously uncomfortable passage to Calcutta in a Malay *prau.*

Minto, educated in Paris, had a wide appreciation of the dangers of French control. Trafalgar had eliminated the threat of a naval blockade of the Straits, but 1812 had not begun and Waterloo was four years away, and Napoleon still nurtured an elaborate plan for an overland attack on India and the conquest of a new empire in the East. Raffles left Calcutta charged to organize the capture of Java from the French.

Precisely the tool that Minto required, Minto was precisely the overlord that Raffles needed. Java was taken in a campaign that, for its era, is a masterpiece of timing, of organization and of impetus, and Raffles's reward was his appointment as Lieutenant-Governor of the new territory.

It lasted four and a half years – for almost as long, that is, as Java remained in British hands. When in 1816 it was actually handed back to the Dutch, however, Raffles had already left under vague charges of maladministration. An acrimonious quarrel with his military commander and the spite of a secretary dismissed for peculation had coalesced into formal accusations of dishonesty in the purchase of land. He returned to London with the enquiry still undecided.

Astonishingly, he was at once accepted at Court. A correspondence with the young Princess Charlotte developed into a remarkable friendship, and he was knighted at a levée held to pay tribute to his monumental history of Java, 'the greater part of which', said the Prince Regent, bestowing the accolade, 'I have read.'

Raffles's first wife had died in Java. In 1817 he married again, took up a provisional appointment to the insignificant Presidency of Bencoolen on the south-east coast of Sumatra, and sailed. Before he left he lodged with George Canning, Chairman of the Board of Control, a memorandum laying down the imperative necessity for the establishment of a base at the deep-water exit from the Straits of Malacca. Napoleon's biting comment on the handing back of Java – 'Probably for a thousand years such another opportunity for aggrandizing England will not occur' – was already current in England.

A year later Raffles went back to Calcutta again with a fresh and urgent plan.

The Marquis of Hastings, with whom he had had trouble in the past, was Governor-General now, but for a second time Raffles came away with the sanctions that he needed. Three groups of islands spread across the exit from the Straits of Malacca – the Rhio archipelago, twenty miles off the southern tip of Malaya; the Lingga group to the south-west; and the Carimons in the gullet of the Strait itself. Raffles was authorized to attempt the acquisition of a base in one of these. And at the ultimate moment, as his ship waited for a wind, Hastings empowered him in case of need to try the Sultanate of Johore itself.

He reached Penang on the last day of 1818 to be told, acidly, by Colonel Bannerman, Lieutenant-Governor of the Presidency, that the Dutch had already seized the Rhios, had probably secured the Linggas, and were in *de facto* control of the Straits.

Raffles had a talent for making friends in high places, he had an equivalent facility for making enemies at his own level, as Java had demonstrated. Bannerman had aspired to action in the mouth of the Straits himself but had pressed it with inadequate enthusiasm and had failed to secure support. He resented both Raffles's energy and his authorization, and determined furiously on obstructionism.

William Farquhar, Resident at Malacca, had originally suggested the importance of a post in the exit from the Straits to Bannerman. His plan had provided for the seizure of the Carimons. Now, in the face of Bannerman's obstructionism, he threw in his lot with Raffles. When, on the grounds of the danger of a quarrel with the Dutch, Bannerman refused troops for the expedition despite Hastings's orders, Farquhar agreed an elaborate cover plan with Raffles. While Raffles, ostensibly waiting for a decision from Calcutta on the dispute between himself and Bannerman, remained behind, Farquhar sailed with one frigate, a small number of country ships, and a grudgingly agreed escort of sepoys on a reconnaissance mission to determine the actual situation in the mouth of the Straits.

Raffles slept one night more in Penang and sailed at dawn in the Company's ship *Indiana* (of whose readiness Bannerman was not aware), leaving a message that he considered it necessary to supervise Farquhar. Ten days later he caught up with him off the Carimons.

On 29 January, with Farquhar and one sepoy, he landed at Singapore village and presented himself at the house of the Temenggong, the Chief of the island.

The Temenggong, a minor member of the Royal House of Johore, was altogether disposed to make a deal; there were, however, problems. Lying deeply indented in the mainland of Johore, the island was an integral part of the Sultanate. The Sultan had died six years previously. In the absence of his elder son – the Tunku Long – the younger had assumed the regency and, backed by his uncle in association with the Dutch, refused to surrender it. In exchange for recognition, the Dutch had been granted rights both in the Rhios and in the Linggas. The Tunku Long had acquiesced in the arrangement, for lack of support.

Raffles decided instantly that the situation demanded an urgent exercise in king-making. Discovering the present whereabouts of the Tunku Long, he sent native craft to fetch him and began, with a balanced recognition of priorities, work on a fort of twelve guns and preparations for an elaborate ceremonial.

On the first day of February the Tunku Long arrived with a not altogether unnatural apprehension in regard to the Dutch. Wisely, Raffles put the Temenggong, who had a fine fat rental for the island in prospect, to work on the Tunku, in conjunction with another member of the ruling family. With Raffles's own oratory, his very real knowledge of local custom and language, and the vast prestige that had accrued to him in Java, the thing was done.

On 6 February, with the ships dressed over all, thirty naval officers in full uniform for personal support and a guard of honour of just 340 sepoys, Raffles installed the Tunku Long as Sultan of Johore. A hastily drawn up treaty was sealed and signed, salutes were fired, and the Union flag was hoisted. Singapore changed into an appendage of the British Crown, Farquhar was appointed Resident and Commandant, and the Temenggong got his cut.

On 7 February, having in ten days conceived, secured and systematized one of the major cities of the nineteenth century, Raffles, leaving

Farquhar with an enormous sheaf of directives, sailed for Penang and battle with the infuriated Bannerman. It was his contention that the Dutch would scream in protest and do nothing. It was his conviction that he would have his real trouble with his own associates.

He was right on both counts. But Hastings backed him, as he had promised, the commercial community at Calcutta hailed the exploit with enthusiasm, and in May Bannerman was dealt with firmly from above. Late in the month Raffles took his wife and new-born son back to Singapore. In three and a half months the somnolent fishing village had expanded into a town of 5,000 people, largely Chinese from Malacca and Rhio but including a reasonable proportion of Malays. Farquhar was instructed to bridge the river that made the harbour, lay out a township with roads, plot housing sites, set up watering places for the ships, and build a respectable residence for himself. The Temenggong was made a member of the council.

Raffles never did things by halves.

Three weeks afterwards he sailed for Bencoolen by way of the Strait of Sunda. He described it as 'a scene of enchantment'. Krakatoa was still quiescent.

Raffles went back once more. In October 1822, desolate, three of his children dead in tragic succession and his wife and himself atrociously ill, he landed at 'the large and prosperous town of Singapore'.

The population had grown already to more than 10,000 souls. Four hundred European ships had used the harbour in its first two and a half years, together with 2,500 country-built vessels; trade had passed the 8,000,000 dollar mark – more than the total of Malacca and Penang together. Singapore was established.

Its success did not prevent the now almost inevitable conflict with Farquhar, who could not think largely enough for Raffles's eager mind. Farquhar was relieved of his post, and Raffles returned to Bencoolen where he and his wife suffered an endless succession of attacks of tropical disease and where, finally, they lost a fourth child. In July he applied to return to England.

No ship was available until the proposed sailing of the East Indiaman *Fame* the following January. He waited with an increasing desperation, gathering together compulsively the last of his great collections of books and Oriental manuscripts and native art, of drawings and paintings of natural history, of plants and animals – stuffed and live.

Early in January, with the arrival of the *Fame* still problematical, a small country-built brig anchored in the port. She was cramped and inconvenient, but Raffles was by now past caring and, in addition, he knew her people – she was the brig *Borneo*, belonging to Alexander Hare and the Ross brothers. Had he sailed with her, he would have gone to the Cocos Keelings on the way home, and a detached and scientific view of Hare's singular story might have become available.

At the last moment the *Fame* was sighted. Raffles reverted to the original plan. Everything that he possessed, his botanical, geological and artistic collections were loaded; the wild animals – the new species of tiger, the first tapir to be sent to Europe, the orang-utans and the monkeys and the birds that were to form the nucleus for the London Zoo which he and Sir Josiah Banks proposed to create – were sent on board.

Fifty miles out in the darkness of the first night the *Fame* caught fire above a cargo of saltpetre. Raffles lost everything, even the great diamond that the Princess Charlotte had given him. He reached London shattered in health.

Farquhar had also returned to London and was claiming acrimoniously that the concept of Singapore was his alone. The eventual rejection of his claim by the Court of the East India Company gave him cold comfort. He had in reality made significant contributions to its origins and he had served faithfully in the early stages of its settlement.

It gave Raffles small comfort also. Already old at forty-five, his life was nearing its end. In 1826, seven years after that hot and colourful ceremony at Singapore, he died – dunned by the East India Company for salary drawn on leave in England, for losses due to the depreciation of the rates of exchange, for expenses incurred in the expedition to Singapore.

After his death the Court extracted £10,000 from his widow.

The impact of World War II on Malaysia is a classic lesson in the dangers of the overstretch of empire. Like Greece, like Rome, like Renaissance Spain, Britain had acquired too much.

Hong Kong, Malaya, Borneo, New Guinea, the Solomons, and the western archipelagoes of the Pacific were added to the Crown together with rights to the Treaty Ports in the nineteenth century in the belief that they would always be protected by the Royal Navy.

At the time of their acquisition the thesis was acceptable. Its weakness was that it failed to take into account the possibility of the rise of a major naval power in the East.

Japan's emergence in that role after the Russo-Japanese War of 1905 altered at once the history of Asia. Japanese naval ambitions made possible continental conquests, military emulation made them obligatory. The process which began in Korea extended itself to Manchuria, proclaimed its ultimate purpose at Shanghai and spread, with World War II, across South China and into Indo-China until it engulfed Malaya and Indonesia. It was one process, interrupted only by illogical intervals of peace, and its ultimate aim was the establishment of The Greater East Asia Co-Prosperity Sphere, dominated by Japan.

In the end, Japan overstretched her imperialism more swiftly, more drastically, and more disastrously than any of the nations before her, but in the execution of her plan Singapore was overwhelmed.

The Malayan campaign is one of the distressful episodes in British history. Criticism of it has been for the most part ill directed. That there were local failures and individual incompetences is, of course, plain. That anything – or anyone – could, in the circumstances of the middle phase of World War II, have saved it is not conceivable. Singapore was lost not when the 15-inch gun emplacements which absorbed so much of the funds for the defence system were built to fire out only to seaward – that, after all, was their essential purpose – but when China collapsed against the Japanese 'incidents', when France fell, when Indo-China crumbled, when *Prince of Wales* and *Repulse* were sunk because the battleship had outlived its usefulness, when the demands of the European war made air reinforcement impossible, when Pearl Harbour changed, for a significant period, the balance of power in the Pacific.

In the time scale of the war in Europe, and of the process of American revival, it was inevitable that Malaysia would fall – and Singapore with it.

Singapore revived, the inherent principles behind the original concept were sounder than the concepts of its defence. It has today a population of more than 2,000,000. One and a half million of these are Chinese, the remainder Malays, Indians, Europeans, Eurasians. Thirty thousand ships a year use its harbours. Its annual revenue is in the region of

£133,000,000 sterling. It exports about £740,000,000-worth of goods a year and it imports £620,000,000-worth. It handles the largest entrepôt trade in South-East Asia.

No longer a fragment of the Sultanate of Johore, it is also no longer a colony. In 1965 Singapore separated from Malaysia to become an Independent Sovereign State, tied by bonds of common interest to the Commonwealth, justly proud of a century and a half of prodigious accomplishment.

It still remembers Stamford Raffles.

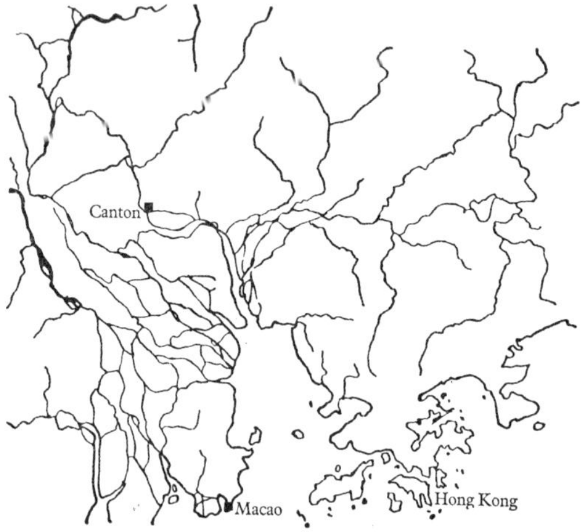

MACAO–SHAMEEN–HONGKONG

In the all but indivisible co-operation of God and Mammon that characterizes the upper levels of Renaissance expansionism, Macao – the Loyal City of the Name of God – was founded in 1557 at the mouth of the Pearl River.

It was a logical development, following almost inevitably on the breakthrough into the Indian Ocean by Bartholomeu Dias seventy years before. India was tapped first, the eastern side of the Indian Ocean was explored early, the gates through the island barrier of Indonesia were discovered, and the route to China was opened.

For the first fifty years trade was tentative and exploratory. China was still as exclusive as it had become in the days of the collapse of the Mongol dynasty. Getting there first, Portugal made vigorous efforts to match that exclusiveness. Britain in swift pursuit made parallel efforts to break her precarious monopoly. A permanent establishment in China was, to the Portuguese, clearly essential to protect their interest.

The peninsula of Macao, a six-mile sliver of high ground pendant from China on the narrow causeway that is known as The Stalk of the Lotus, fulfilled the Phoenician requirements. It was defensible – in terms of the age. It tapped a vast hinterland – in terms of trade. It was a focal point – in terms of communications.

The Si Kiang river is navigable to shallow-draught vessels for 1,000 miles across South China. It is the main artery of transportation for a vast and productive population, and its traffic has been systematized for perhaps 3,000 years. Close to the sea it divides into a complex delta. The Pearl River is the northernmost and the most serviceable channel through the delta. Canton – almost certainly Cattigara of the *Sinae*, the ultimate goal of the classical geographers – lies on it at the crossing of the main routes of communication with Pekin and the North.

Macao huddles behind a maze of islands at its point of junction with the sea. Precisely how the peninsula became Portuguese has never been made entirely clear, but a tradition of corruption among officials sufficiently far removed from Pekin to be quasi-independent is early apparent in the immemorial custom of the East. Portugal did, in fact, pay rent – or alternatively tribute – for Macao for nearly 300 years, and total sovereignty was not recognized by treaty until 1884.

Notwithstanding the rent, Macao showed a profit early in its history. It continued, despite the *dolce far niente* it imported from its homeland (once the great years of the exploration were over and Portugal had lost her drive), to provide a substantial share of the income of the nation. It also remained singularly Portuguese. In this perhaps it was assisted by the fact that for its first 130 years Chinese were forbidden to live on the peninsula, but possibly also because God for a time had the measure of Mammon – Macao had a Bishop a century before it had a Governor! Even the arrangement by which the East India Company shared the peninsula for a number of years failed to alter its national characteristics.

Through the squabbling of the eighteenth century and the upheavals of the nineteenth Macao survived, effective rather than efficient, adequate rather than ambitious. Always a little apprehensive because the Pearl River was silting up the approaches to its less than perfect harbour, always a little threatened by Chinese nationalisms.

In the gathering years of World War II it was quietly prosperous, the unobtrusive partner in the triumvirate of extra-national entities that divided between themselves the rich cream of the trade of Southern China.

The Shameen was, in point of history, the second partner.

The Shameen is a sand-bank island – a mud-bank rather, that by the caprice of the Pearl River fronts a long sector of the City of Canton. A channel divides it from the City. In 1861, when the military occupation of Canton was lifted after the Second Opium War, it became an international trading quarter.

Though intermittent and at times surreptitious trading with Canton by other nations began almost as early as the Portuguese trade, it was not until a century and a quarter later that the East India Company broke the Portuguese monopoly and secured a solid foothold in Canton itself. Fifty years later the French followed, and almost fifty years after that, the Dutch.

Chinese policy had changed, the economic possibilities of the export of tea, silks and china, and oddly enough rhubarb, were beginning to be appreciated. Canton was declared sole entrepôt for international trade and the East India Company was granted rights in an area under a factor. In the earlier period at least, merchants, British and others alike, lived within the 'factor-y'. The new regulations laid down that they traded through the *co-hong*, an association of Chinese merchants accredited by Pekin and subject to stringent regulations.

For a century and a half the system operated with a relative efficiency. The number of factories expanded, the number of nations increased, the number of traders multiplied, the opportunities for graft also expanded, profits were maintained despite problems.

First among the problems was opium.

As early as 1783 Warren Hastings had declared, with that talent for the *non sequitur* that earned for British statesmen a reputation for hypocrisy, that opium was:

> A pernicious article of luxury which ought not to be permitted but for the purpose of foreign commerce only.

Shedding a manly tear, the East India Company itself declared, a few years later, that:

> If it were possible to prevent the use of the drug altogether, except strictly for the purpose of medicine, we would gladly do it in compassion to mankind.

The root of the subsequent trouble was that China was at least equally ambivalent. China deplored the consequences of opium, made increasing efforts to prohibit its import, resented the balance of payments problems that it caused, closed her eyes to the monstrous corruption of the *hongs*, and permitted the cultivation of vast tracts of opium poppies. That much of this stemmed from the corruption of her own officials is irrelevant. She permitted the corruption.

Eventually and very admirably she determined on action. In 1839 a commissioner was appointed. With immense energy stocks of opium in the European factories were seized for destruction, bonds against the future importation of the drug were demanded and pressure became ruthless.

The East India Company might have managed the affair tactfully, but the East India Company had been forced seven years previously by rival merchants and political pressures in England to surrender its monopoly in the China Trade – the East was always an intemperate breeder of monopolies! Britain went to war instead.

It was on the whole an interesting campaign. It proved Britain's contention that a properly organized navy could bring properly organized pressures to bear anywhere in the world. More locally, it investigated an excellent deep-water anchorage between the peninsula of Kowloon and an offshore island. It established proper navigational systems in the area, and its land force, progressing slowly and comfortably up country, provided excellent training for the troops as far as the Yangtze, where they threatened communications between Pekin and the south. The Imperial authorities decided on compromise.

In terms of the subsequent Peace Treaty, five ports were thrown open to international trade, the *co-hong* was abolished, a fair and regular tariff was agreed, Britain obtained 'most favoured nation' treatment, and Mr Caleb Cushing negotiated an excellent treaty for America. The island which protected the anchorage of Kowloon was, at the suggestion of the Imperial commissioner, ceded to Britain – its name was Hongkong.

Britain, none the less, declined to give guarantees about the opium trade for the future, the Chinese declined to make the trade legal, and twelve years later it was all to do again.

This time the French joined with Britain – which was just as well, for the Indian Mutiny was about to occupy much of the interest of the British Army.

In the Second Treaty a fixed tariff for the importation of opium was agreed and the trade legalized. Simultaneously, the activities of Christian missionaries were sanctioned. Other concessions were in due course agreed, and when the road of the envoys to the ceremony of the ratification of the treaty was barred, the allies fought their way through to Pekin and burned what had been left by revolutionary Chinese troops of the Summer Palace of the Emperor – who had very sensibly removed himself to Jehol.

At Canton the merchants, now acutely aware by experience of the vulnerability of extra-territorial enclosures in potentially hostile towns, reclaimed Shameen and built along it the most astonishing congeries of mansions, offices, counting houses, warehouses, and clubs, all in the middle style of Good Queen Victoria – with Oriental embellishments!

Oddly enough, the Shameen, despite its architecture, served its purpose until in the inextricable confusion of nationalism and the Japanese invasions, the fevered tempers of the Opium Wars were reborn in the shadow of World War II.

Hongkong, the third partner of the triumvirate, came into existence by the First Treaty in 1842.

It already shared an interesting association with the opium trade. From 1821 ships of the trade had been in the habit of sheltering in its capacious harbour, and it was their masters who indicated to the Royal Navy the advantage of the anchorage.

It never looked back.

Its vital advantage over the other members of the triumvirate was that it alone fulfilled, precisely, all the Phoenician specifications. The Shameen was, it is true, an island of a kind, and it was therefore possibly immune from mob violence. It was not, however, in any sense defensible in normal military terms and it was, in addition, distantly removed from the trade routes and the open sea.

Macao was perhaps more nearly defensible in the military sense, but already its harbour was silting up and its connection with the main traffic routes growing uncertain.

Hongkong, on the other hand, stood in deep water. It had superb entrances east and west. It could command the Pearl River with minimum force. Its anchorage was perfect and it was large enough to support an interior life.

It was, however, even smaller than Singapore. Barely eleven miles long, its total area was – is – thirty-two square miles. Rather more than eighty per cent of this is pre-empted by The Peak, the 1,800-foot mountain whose summit commands a superb panorama of the coast of China in which islands float away, like swans retreating, in an infinite perspective of beauty.

Hongkong began its life as an opium port.

By 1849 it had reformed. It was exporting live-stock, human live-stock – coolies to the goldfields of San Francisco and Australia. It was

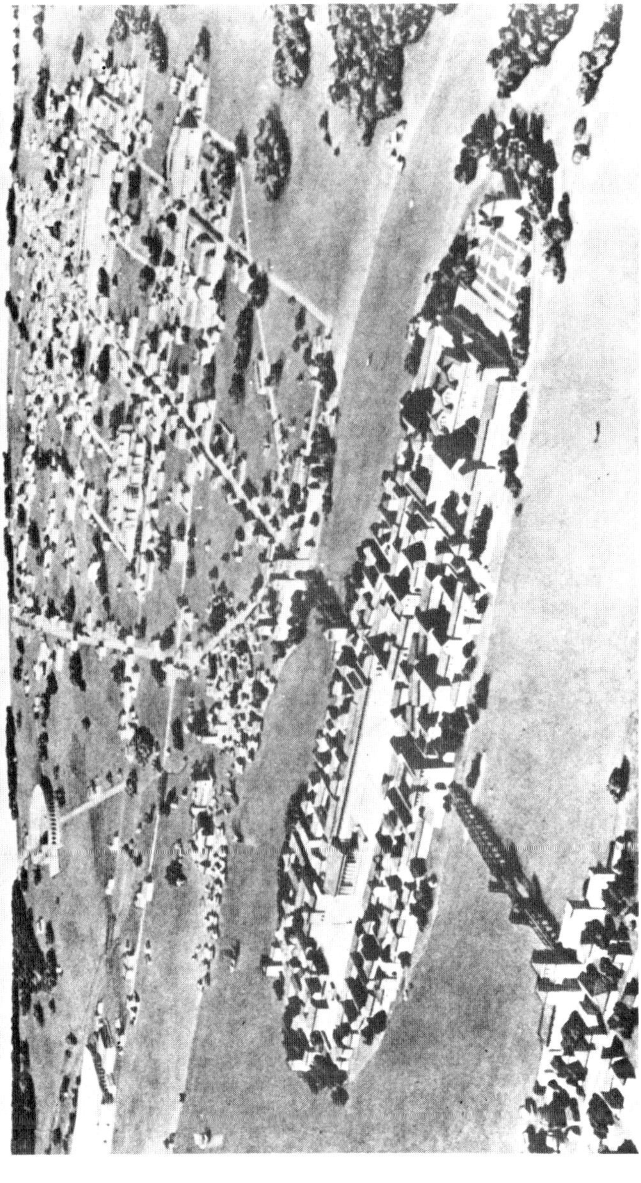

'My dear Lutetia . . .' The exquisite Roman city that once rode the Seine like a ship at anchor.

(Opposite p. 185)

Singapore in 1840 – the city island of two million people that was conceived, secured and systematized in ten days by Stamford Raffles.

only after this phase, in 1850, that it began to approach the problem of providing a Chinese population for itself.

With the end of the Second Opium War it acquired Kowloon, the peninsula that juts into the harbour a mile across the water and that might conceivably menace it in war. About the end of the century, for greater security still, it leased the New Territories, as far north as the Shumchun River.

The lease expires in 1997.

Growth in Hongkong was remarkable, even by twentieth century standards. At the end of its first fifty years it drew level in population with Macao. It moved rapidly to the million mark in 1936. By 1941, as its centenary year coincided with the Japanese attack, it counted 1,729,000, 150,000 of whom lived in the vast flotillas of junks and sampans.

The campaigns against Singapore and Hongkong are utterly dissimilar, the military philosophies behind them identical, results were the same.

Hongkong, like Singapore, was conceived in the belief that a strong navy could render it inviolate. In the days of sail the contention was unquestionably true. In the days of the aircraft carrier, the bomber and the tank it was an illusion.

As early as 1921 it was officially recognized as such. In that year the Committee of Imperial Defence arrived at the conclusion that Britain's only naval base in the Far East

> . . . was too close to Japanese naval and air bases in Formosa and too remote from the nearest source of British reinforcement to be strategically sound.

'There was', says the Official History, 'no possibility of making Hongkong secure against attack.' It was on the basis of this decision that the plan for a major naval base at Singapore was adopted. Twenty years later it too was strategically unsound.

Hongkong began in reality to live on borrowed time when the Japanese invaded South China at Amoy in May 1938. That landing carried, implicitly, the threat of future disaster. By October, a little after Munich, strong forces landed in Bias Bay, the notorious pirate area thirty-five miles east of the Colony. On 21 October the Japanese, moving with great speed across hill country, captured Canton and severed connections between the Chinese Nationalist armies and Hongkong.

For a year there was relative inaction, though Hainan Island to the south was taken in February 1939, but when on 7 December the traumatic attack on Pearl Harbour broke it was clear that little time was left.

Early on the following morning Kai Tak airfield on the Kowloon side was bombed.

The defence of Hongkong is a sad little story of unjustifiable optimisms, understandable misjudgements, great courage and self-sacrifice. It failed. It had to fail.

Hongkong was the second of the triumvirate to go.

Incredibly, despite Japan's basic attitudes towards the European enclaves in the East, and in disregard of the fanatical right-wing parties, Macao was permitted to exist, unoccupied. It was harassed from time to time, threatened endlessly, squeezed, and its islets were at one period taken over. Its only serious damage, however, was from an American carrier attack. The attack was in error.

The effects of war to an observer in the first days of peace are necessarily subjective. In the great triangle of the delta of the Pearl River there was less tangible destruction than almost anywhere in the East. Nothing in any way resembled the burned out wasteland of the Tokyo fires, there was nothing like Yokohama – nothing remotely like Hiroshima.

I reached Hongkong in a destroyer from Japan in the morning of 16 September, two hours before the formal ceremony of the surrender.

Hongkong was vitiated. No great damage was apparent. What there had been was cleared or patched over. But it lacked paint, there were cracks and dislodged plaster, windows were boarded up. It had the air of something from which the life had been drained, slowly, and over a long time. I remember the aching desolation of Stanley Prison and the houses picked clean on The Peak with windows like eyeholes in a broken skull. I remember the mummified body of a forgotten child in Kowloon. It was despairing.

The surrender was quietly ceremonial. General Okada and Admiral Fujita signed the articles impassively. Their slender swords were yielded to Admiral Harcourt. There were no drums, no trumpets.

Afterwards, when I went down the great staircase of Government

House, I heard a senior officer say: 'And they've even fitted out the rooms upstairs with tatami matting and sliding screens. It's like – dammit, it's like a Japanese brothel!'

I crossed the next day to Macao. It too was a little worn, but with the ordinary and characteristic shabbiness of the East. It smelt strongly, but it always had. It was overcrowded, though some of the refugees that it had taken in from Hongkong had gone already – and more were going. It pulsated with life.

The Central Hotel was then at the splendid summit of its wickedness. After twenty-five years I have a confused memory of a vast banquet – everybody was celebrating peace – and the fan-tan rooms with the baskets coming down from the balcony, and the constant noisy bustle. There were nine floors: gambling floors, woman floors, possibly opium floors – it was never very clear. There was always an involuted complexity of movement, a swirling, rococo life.

In the daytime it was strangely honest: the hieratic junks moving majestical and slow across the placid water, the sampans weaving a pattern through them, the long wharves of the fishermen with nets, and the rope-makers, and the men repairing sails. The pace of movement was not neurotic, as in some other parts of China, but it was ceaseless, unending.

I remember that at the banquet was the famous Fu Tack Yam – 'Mr Fu', who owned the Central then, who owned the gambling concession, who owned restaurants and tea houses and dance halls, and who lived in a house with sunken gardens and barbed-wire fences, and smoked in a pavilion dedicated to a notable pirate of Bias Bay . . . his father!

Macao had changed a little in the course of the war. It was, the Old China Hands said, wickeder.

I left it in a still evening when the water was like pale mother-of-pearl in a junk with a motor like a traction engine. It was an offence upon the water, and I slept in a cage. I had no map – I do not think that the junk had a chart. I had no Chinese. But I believed I was going to Canton.

It was a strange, broken journey. I transferred to a smaller junk with sails, and we were fired on from the pirate villages near Boca Tigris (not, I think, purposefully but out of custom). I was transferred to the shore and moved on the pillion of a bicycle, with a sweating youth to do the pedalling. I remember sleeping on the steps of a temple between statues of gods, and sleeping in a camp among soldiers, and sleeping in a delightful white house with a window like a moon. Afterwards I arrived in Canton in a sampan. I never asked how. It did not seem to matter.

The afternoon of the day that I arrived I found my way to the Shameen.

It was altogether dead.

For a long time I was quite alone walking up the path that was the main road, past the shuttered Victorian houses and the warehouses that were falling down. Once or twice I saw men moving round corners, but there seemed to be no one living there, no indications of activity. Paintwork was peeling; where the windows showed, they were broken.

When I had seen enough, I turned back.

Later I saw soldiers on the path. It was some time before I was certain that they were American. I was not prepared for them. They were, I gathered afterwards, an advance party coming down from Chungking. I was bemused at the time. When I had passed them a Major came walking sharply back and asked me why I had not saluted the General. It was quite impossible to think of a reason. I said simply that I didn't know, and he considered this and, after a moment, appeared to accept it as rational, and we separated. It was all entirely friendly. I think that this, possibly, was the moment that I realized that the war was over.

Whether in other circumstances the Shameen could or would have survived is, naturally, debatable. Under a Communist regime which rapidly elevated the exclusiveness of the Mongols to the point of religious fanaticism, it was beyond reality.

The triumvirate diminished to two.

Macao burgeoned – precariously. Mr Fu was kidnapped and a slice of his ear was sent back to his loving family with a demand for 1,000,000 dollars, but gambling barely missed a throw. Vice achieved new and more ingenious summits and began to diminish. Finally, in 1948 Macao discovered gold. It was not the traditional discovery of the raw metal in river sands or in veins of quartz, but the discovery of the possibilities of a stream of properly worked bar gold, flown in by aircraft without passengers (one of which was destroyed in the very first of all air hi-jacks) or delivered by sea. Its entry was entirely legal; there was a businesslike Gold Syndicate whose members were known, there was even a profitable and increasable tax. It was its departure that was mysterious. Even now the destination of much of it, by ingeniously covered channels into countries whose laws prohibit it, is impossible to discover. With it, other illegalities built up.

Ian Fleming was instructed in its possibilities on a visit to Macao and based, to the sardonic amusement of Hongkong, some of James Bond's more improbable adventures on a romanticized view of it. Most of the fantastic facets of the Gold Trade he attributed to the remarkable Mr Lobo of Villa Verde. This appears to have been erroneous; Old China Hands say that from the start it was masterminded by Mr Ho Hyn – or Yin, the spellings vary – one time partner of Mr Fu, long time Chairman of the Chamber of Commerce of Macao and, as a respected member of the Political Consultative Committee of Pekin, all time 'running dog of the Communists'.

To what degree the immunities of Macao may be attributed to Mr Ho Hyn is not, of course, admitted by either side. Certainly the troubles of 1961 and 1963 were smoothed over, and the bigger troubles of 1967 were pragmatically adjusted. After each of these ebullitions the degree of real Communist control in Macao beyond question tightened; it was *not*, however, taken over. Macao survives.

Hongkong pursued a wider goal. In the years after 1945 it endured a brief uncertainty, adjusted its views and its circumstances to a new environment, and returned to life with the vigour and the excitement of the Japanese 'flowers' that as a child one dropped in water and watched expanding until they filled a young imagination.

Courageously the consequences of the Japanese occupation were faced, the deterioration due to war was arrested, the refugee problem was mastered. New markets replaced lost ones, old markets where possible were revived. New industries came into being to balance losses in the entrepôt trade and to absorb the successive waves of refugees. A million people were rehoused. Hundreds of factories were built for light industry. Skyscrapers rose first along the front of the City, then in Kowloon, then as far as the eye could stretch, almost to the Lei u Mun Pass. The population rose to 4,000,000.

The regeneration of Hongkong is an economic miracle, it is at the same time a political miracle. The two things are intimately associated.

Hongkong's own revenue today tops £135,000,000 a year. Its imports total £780,000,000, its exports £500,000,000. It is worth at a conservative estimate £250,000,000 a year to Communist China in foreign exchange. Two-thirds of that comes from direct trading, a third from remittances.

Hongkong, according to the old China Hands, is Communist China's gateway on the outer world.

Macao, the cynics say, is China's window on the underworld.

Early in 1969 there was a new Communist slogan: 'The Chinese people will present Hongkong as a humble gift to Chairman Mao in 1970.' The date can, apparently, be read alternatively as 'in the seventies'. It should be considered together with the flow of foreign exchange, and the symbolic importance of the fact that the skyscraper of the Communist Bank of China overtops the skyscraper of the Hongkong and Shanghai Bank by twenty-five feet, in any attempt to prognosticate the future of the colony.

Piracy

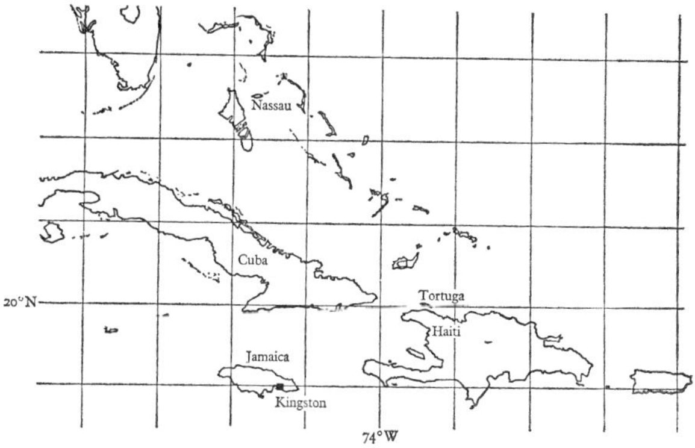

THE WEST INDIES

These last islands swing in a long lazy curve like that of a tired ostrich feather across 2,000 miles of sea from Cabo San Antonio at the western point of Cuba to Trinidad off the mouths of the Orinoco. Columbus found them and, as has been told, Columbus, at Navidad, brought evil into Eden.

Good or evil, they are unutterably beautiful. High green islands for the most part, sculptural and lovely: Cuba – that brought the world to the edge of nuclear war a decade ago; Jamaica – of the buccaneers; Haiti – and Voodoo and the Black Emperors; Puerto Rico – where legend says that the Spaniards exterminated 600,000 in ninety years; Anguilla – that made a mock of a British Cabinet a brief while back; Martinique – and Mont Pelée that almost outreached Krakatoa; St Lucia and Barbados and Trinidad.

Romance has many facets.

The West Indies, in all their astonishing variety, are, none the less, strung together on one common chain – the chain of Piracy. They are the lawless islands – the islands where the Privateers achieved their most tremendous triumphs, the islands of the Boucan men, the islands of the Pirate Republic. Nowhere else in the Seven Seas did all three of the great divisions of sea crime interlock.

Privateering, as far as Britain is concerned, goes back 700 years. Edward I issued 'commissions of reprisal' to ship owners plundered in war or peace, sanctioning them to plunder ships of the same flag as their plunderers. In a turbulent age it had a cool logic.

Over the intervening years it was said of most privateersmen that they were 'pirates with a piece of paper'. Intermittently that paper was recognized. At other times they were hanged like the rest, on capture.

The English privateeers were not the first among the islands. The French hold that honour. Rivalry between France and Spain at the level of the Kings loosed in the early 1500s a host of 'corsairs' on them, brawls with the Dutch over the salt trade later in the century let in the angry Lowlanders. The first British privateer, it is true, had been recorded – but not named – by Alonso de Ojeda off Maracaibo in 1499 but after that, identifications are rare until John Hawkins went in in 1562.

Hawkins sailed with three small ships and a renegade Spanish pilot to the West African coast. Whether he carried 'commissions of reprisal' or 'letters of marque' is entirely doubtful. That he went with the knowledge of Queen Elizabeth is almost certain. That his attack on the Portuguese slavers of the Guinea coast was simple 'hijacking' has already been discussed. By it he

> . . . got into his possession, partly by the sword, and partly by other means, to the number of 300 negroes at the least. . . . With this praye he sailed over the Ocean Sea unto the Island of Hispaniola.

The second part of his venture also has an apt description in modern terms. It was simple 'muscling in'. He forced his way inside the Spanish enclave, bullied and bribed the authorities of Isabella, sold his negroes there and at other ports, and loaded his ships under duress with hides, ginger, sugar, gold, silver and pearls.

Admittedly hides were the main cargo but the venture cleared a handsome profit, though two ships which he sent, with a consummate impudence, to Spain, were promptly seized. It is at least possible that he planned their seizure to provide justification for a second voyage.

Two years later he anchored off Margarita Island on the Spanish Main. His status had altered. At Borburata he wrote to the Governor demanding harbourage and repairs.

> Whereby by order of Elizabeth, Queen of England, my mistress, whose fleet this is, I cleared on a certain voyage, and was by contrary weather driven to these coasts. . . .

It wasn't her fleet and the weather had been fair. The kernel of the matter was that he felt strong enough to say:

> If this petition be not granted, I shall seek my own solution.

The Governor hesitated. Hawkins landed his small force. The Governor capitulated and gave hostages for good behaviour.

Whether the Queen had a share in the joint stock company that backed the second venture is uncertain, but when the Spanish Ambassador made strong protests at court against a third voyage that Hawkins was preparing, the Queen forbade him to go. Hawkins at once sent his ships secretly to sea under John Lovel, with the young Drake attending him, and stayed amiably at home.

The voyage was a failure. Inevitably Hawkins sailed again with the fourth venture in 1567; this time he had two Royal Ships, the *Jesus of Lubeck* and the *Minion*, flying the Royal Flag. Drake sailed in command of a third ship, the *Judith*. Again Hawkins landed under threat and made his deals, adding bribery to shows of force. At Cartagena, however, he was refused. It was the strongest fortification in the Caribbean. Hawkins bowed at last to the impossible and set course for the Yucatan channel.

From Cabo San Antonio on Cuba, at the very point of the beginning of the islands, he was forced back into the Gulf of Mexico. His ships suffered weather damage, but eventually he reached the rocky islet of San Juan de Ulúa to seaward of the city of Vera Cruz in Mexico, having seized, and subsequently released, three small Spanish vessels on the way.

The Spaniards, taking him to be part of the expected Plate Fleet, let him in, and he seized the islet and its defences.

Work began at once.

Exactly one day after, thirteen galleons of the Plate Fleet stood in from Nombre de Dios. It has an intense dramatic impact. Hawkins might have held them off, but the Queen had put sharp limits to his belligerence. He negotiated. Hostages were exchanged and when three days had elapsed the Spaniards were permitted to enter their own harbour. Inevitably they made a 'treacherous' attack. Hawkins, a survivor said afterwards:

> . . . with a loude and fierce voyce called unto us, saying, God and St George, upon these traiterous villaines.

The *Jesus of Lubeck* was dismasted, the Spaniards seized the islet and the batteries, fire ships were sent in. Only the *Minion*, to which Hawkins had transferred, and the *Judith*, with Drake, got clear, and '. . . the barke the same night forsook us in our great miserie'.

None the less, they survived. Ruthlessly Hawkins put half his men ashore on the Mexican coast to ensure that the food would last. In January of the following year he limped in to the English Channel with many of his men at the point of death. Drake had been in for five days.

Hawkins rose to high office in the Admiralty. As Treasurer to the Navy he nursed the slender fleet through Elizabeth's parsimony and the political uncertainties of the reign. His genius shaped new types of ship and new types of tactics; he was Rear Admiral of the fleet that drove the *Armada Felisisima* up the Channel.

Drake himself returned to the islands while his temper was still high. In 1569 he made the first of his two 'secret' voyages to the Spanish Main. Little detail has emerged in connection with them. He was believed to be planning an assault on a major port. It is known that the first time he had two small ships, the second only one – and that is almost all. But in 1572 he sailed finally from Plymouth with two vessels, the *Pasha* of seventy tons and the *Swan* of thirty, for the small cove that he had named Port Pheasant on the Darien coast to the east of Nombre de Dios.

On arrival he found a message to himself on a metal plate from a friend, John Garret, warning him that the Spaniards were aware of his secret harbour. A few days later another English privateer, James Ranse, sailed in – the Caribbean was growing populous. Drake accepted Ranse, reluctantly, and moved forward to an advanced base close in to Nombre de Dios, which he called the Isles of Pines; there he captured two small Spanish craft and released a number of slaves. In the early dawn of 29 July, with his brother and John Oxenham and seventy men, he launched his attack on Nombre de Dios.

It was audacious and it almost served. Surprised in the half light, the Spaniards bolted. The raiding party drove deep into the town and located a storehouse filled with silver bars, but the Spaniards rallied. Drake, wounded in the leg in the first skirmish, collapsed from loss of blood; despite his protests, his men withdrew to get him back to the ships. The assault failed.

The Isles of Pines were abandoned and the slender force slipped away to another island, Bastimientos, to recuperate. Ranse left, and Drake devised a fresh plan. An escaped slave had given him the routine of the annual movement of the Peru treasure across the Isthmus to Nombre de Dios for shipment. He determined on an ambush, but the treasure was not yet due and in the interim he made a raid on Cartagena and captured four ships, after which he cruised backwards and forwards along the Main, harrying the local trade. Altogether in the course of the venture he intercepted more than 100 vessels.

In February of the new year news of the movement of the treasure reached him and, with a minute force of eighteen of his own men and twenty-five Maroons from the Darien coast, he landed on the Isthmus. He set his ambush across the divide, having first from a treetop had his sight of the Pacific. The ambush was admirably set, but one of his men, drunk, alarmed the Spaniards. The second venture failed.

With an incredible pertinacity, Drake held on. He had lost half his men when he met a French privateer, Captain Têtu. They joined forces at once, another convoy was already moving across the Isthmus. This time the surprise was absolute. The Spaniards were beaten off the treasure. Têtu was wounded and had to be abandoned with three of his men, but the main party withdrew with the gold, buried the silver, and reached the ships.

Drake still required one thing – a suitable vessel for the passage home. Outrageously, he sailed again for Cartagena and captured one. Then, in defiance of the Spanish ships in the harbour, he passed:

> . . . hard by Carthagena, in the sight of all the Fleete, with a flag of Saint George in the maine top of our Fregat, with silke streamers and ancients downe to the water, sayling forward with a large wind.

Before he returned to the Main again he had made his incredible voyage round the world. When he reached England at its end he carried a treasure worth £5,000,000 in modern money.

By the end of the century the track-charts of the privateers criss-cross the Caribbean and the Mexican Gulf like shadowing on the sketches of a child. In 1594 Whiddon had reconnoitred Trinidad and the Orinoco. In the following year Robert Dudley and Amyas Preston, tired of waiting for Raleigh, had gone raiding on their own account along the Main. Raleigh, with four ships, arrived at Trinidad in March, burned the town of St Joseph, and met with little joy in the Orinoco. Popham, who had joined Dudley at Trinidad, helped him take a Spanish ship off Grenada, and the two of them recklessly attempted to sell her at Puerto Rico. George Somers joined Preston, and they captured the island of Coche and from there raided along the Main. Anthony Shirley went to Dominica, carried Margarita, and captured Villa de la Vega

(Spanish Town) in Jamaica. The list is limitless, yet in a sense it ends with the last voyage of Drake and Hawkins in 1595.

With twenty-seven ships, these two sailed once more from England to cut out a dismasted ship of the treasure fleet reported at San Juan in Puerto Rico. Drake insisted on making a side-swipe at Grand Canary as they passed, and forfeited time. They failed and Hawkins died at sea. Drake lost time again in Dominica and Guadeloupe, and the Spaniards had sunk block-ships at the harbour entrance before he reached it. The defences were dangerously strong. He pulled out to try his fortune once more on the Main. Rio de la Hacha and Santa Marta were captured, Nombre de Dios was burnt, and then, to seaward of Porto Bello, Drake died. With him died the splendour of the Elizabethan privateers.

JAMAICA

Even before the death of Drake off Porto Bello a new aspect of piracy was taking shape along the islands.

In its origins it had nothing whatever to do with the sea. In an earlier section it was noted that the population of Hispaniola diminished sharply after the discovery of Mexico and Peru as the early settlers left to follow the fascination of gold. The north and west of Hispaniola were depopulated, and the raids of French privateers slowly eliminated the undefended villages that remained.

Simultaneously an imbalance of cattle and hogs developed. Left to run wild, they increased until they became a menace to the settlements that still existed. Men were employed to hunt them for their hides; others joined in, a primitive hide trade evolved, and along the north coast and at the entrance to the Windward Passage French interlopers found their way ashore.

Almost certainly it was the French who adapted the local practice of smoking strips of beef and pork over a grating of green sticks into the status of a trade with ships stocking up for a long passage. The gratings were called *boucans*, following a Brazilian word, and the men of the trade were christened *boucaniers*.

As soon as its profits became measurable, the Spaniards opened the customary campaign against interlopers. No dates are available, but by the end of the century the Spanish pressure was heavy enough to drive some of the *boucaniers* off the coast to the narrow island of Tortuga, close to the northern opening of the Windward Passage. Columbus had given it its name as he beat through the channel to the calm of the Limonade Pass and the wreck of the *Santa Maria*!

Obviously, to reach the island, the *boucaniers* had needed at least open boats. They seem to have built up as swiftly as possible a maritime capability, capturing small Spanish craft and using them to take larger ones. They were joined, after the Spanish attack on St Christopher, by English and by French. Their sea strength expanded. They began to plunder the mainland of Hispaniola and to take a toll of passing shipping. When in 1631 the island was taken – with a highly doubtful legality – under the aegis of the Providence Company, Anthony Hilton, an Englishman, was appointed Governor.

His people were perhaps the toughest group on earth, barbarous, brutal, illiterate, intractable. They were united by a single dominant – hatred of the Spanish. They were so ferocious in their expression of it that the authorities of Santo Domingo launched in 1634 the first of a series of suppressive attacks.

The failure of Cromwell's Western Design, paradoxically, supplied in 1655 the springing point of British expansion in the Caribbean. Misled by optimistic assurances of Spanish weakness in the islands, Cromwell had dispatched a force under General Venables in a fleet commanded by Admiral Penn to seize Hispaniola as the first move in a general take-over of the Spanish empire. It was one of the more lamentable incompetences of the Commonwealth. Failing at the hands of 'a sort of Vagabons that are saved from the gallowes in Spaine and go by the name of Cowkillers', the expedition abandoned ignominiously the planned capture of Santo Domingo.

> God was not pleased to deliver it up unto us though with 9,500 men and 80 saile of great shipps and small vessels, soe that never were men more disapoynted than some of us, nor did the hearts of English men faile them more than in this attempt.

The expedition contented itself with Jamaica, which was all but undefended. Even there, by incompetence, they saddled the country with the guerilla of the Maroons which continued for most of the century.

It has been claimed that the buccaneers helped in the taking of Spanish Town. The evidence is unconvincing. Certainly, however, English buccaneers arrived at Port Royal after one of the seesaw changes of control on Tortuga itself (most of them as prisoners), and letters of marque were equally certainly issued to many of their leaders before the end of the Commonwealth.

Port Royal then was a sandy islet at the entrance to the harbour of Villa de la Vega, the end islet of a chain that followed a coral reef and that was connected subsequently by a causeway with the mainland. The Spaniards called it Cayo de Carena and careened their ships there, the buccaneers corrupted it to Cagway. It was sufficiently far from constituted authority to meet their highly particular requirements. 'Organized' buccaneering from Port Royal was in its first phase essentially local. In the early years of the operation it was mainly confined to raids on shipping passing through the great channels. The first major long-range attacks on the Spaniards were in fact made by the Commonwealth Navy. Christopher Myngs, in Richard Cromwell's time, interpreting liberally Admiral Goodson's order to use his 'best endeavours to seize, surprise and take all vessels belonging to the king of Spain', came back to Port Royal having sacked Campeche, Coro, Cumana, and Puerto Cabello. Loot totted up to £300,000 in the money of the day, and Port Royal seems to have entered its first great debauch on the strength of the cash.

Myngs – Sir Christopher now – made a £1,000,000 raid three years later and was gravely rebuked by Charles II, who said that the planters at Jamaica should keep at home and tend their plantations – but accepted his share of the loot notwithstanding.

The line runs direct from there. Buccaneers – the name was extended to all who sailed with letters of marque from the Governor of Jamaica as well as to those who neglected to bother about the formality – captured Tobago and St Eustatius and Saba from the Dutch, outstripping the Navy in the race for spoil. Captain Barnard took San Thomé, Lewis Scott sacked Campeche in Mexico, John David looted the city of Grenada in canoes and got away with 50,000 pieces of eight.

In 1665 Edward Mansfelt, Admiral of the Buccaneers – they had adopted rank now – took *Sancti Spiritus* in the heart of Cuba, sacked Grenada again, and captured the island of Old Providence off the Moskito Coast.

Mansfelt died, and buccaneering reached its apogee. Henry Morgan, Mansfelt's second-in-command at Old Providence, succeeded him as Admiral. He was the nephew of a late Deputy Governor of Jamaica, Colonel Edward Morgan, who himself had led the attack on Saba and St Eustatius in the Leeward Islands. He was not a cattle hunter, it is doubtful if he ever stripped a hide; but he became the most resolute, the most skilful, the most brutal and the most ruthless of the Brethren of the Coast.

Much ink has been spilled in arguing whether or no he was a pirate. He was not. He never sailed without a 'commission' of the Governor. He was more savage, despite the paper, than most pirates.

In 1668, when he made a masterly attack with twelve ships and 700 men against Puerto del Principe in Cuba, the leading inhabitants were tortured to reveal their cash. Later in the year he stormed Porto Bello, locking his prisoners in the magazine of one of the forts and blowing it up, and afterwards forcing monks and nuns to raise his scaling ladders against the walls of the other fort. Esquemeling, who was present, graphically describes the rape, the murders and the looting – he rather admired Morgan! They came away with 300,000 pieces of eight.

The next year he took the towns in the Gulf of Maracaibo and fought his way out through three big Spanish ships, one of which carried forty guns against his own ship's fourteen. The Spaniards reacted violently against Jamaica but achieved little along the coast.

At the end of the year Morgan sacked Grenada once more, seized San Lorenzo on the Chagres River and, with a ferocious energy, led his men across the Isthmus of Panama – where long before him Drake had contented himself with staring at the new ocean – and sacked the city of Panama. The loot was fantastic, more than 750,000 pieces of eight.

Peace between England and Spain had, in point of fact, been signed in 1670. Modyford, the Governor of Jamaica, was arrested and sent to England and the Tower. Morgan was arrested, but 'he had both Sir Thomas Modyford's and the Council's commission and instructions'. In 1674, time enough having been given for Spain to simmer down, Morgan was knighted and sent back as Lieutenant-Governor.

Thereafter he put down buccaneering with a steady hand. He was not entirely loved for it, but when he died – of drink – he was buried solemnly at Port Royal.

Four years after his death Dr Heath, Rector of Port Royal, wrote that he had been

> . . . at the Church reading prayers . . . to keep up some show of religion among a most ungodly debauched people, and was gone to a place hard by the church where the President of the Council was, to take a glass

The pirate city of Kingston a hundred years ago; innocent after the earthquake that swept Henry Morgan in his coffin out to sea.

Anne Bonny, the female pirate whose career provides one of the gaudiest chapters in the entire history of transvestism.

> of wormwood wine with him, as a whet before dinner. . . . I found the ground rolling, and moving under my feet; upon which I said to him "Lord, Sir, what is this?" He replied being a very brave man "It is an earthquake; be not afraid, it will soon be over." But it increased and we heard the Church and Tower fall, upon which we ran to save ourselves. I quickly lost him and made for Morgan's Fort . . . but as I made towards it, I saw the earth open and swallow up a multitude of people; and the sca mounting in upon us over the fortifications.

The church went, a third of the houses of Port Royal went, half the land of the town sank into the sea, and as the tidal wave swept the anchored vessels out through the harbour mouth, the graveyard opened and the coffin of Henry Morgan, Knight, went drifting out to sea.

Port Royal was assoiled.

NEW PROVIDENCE

If the Princes of the Sea who are mentioned in the Bible were pirates, the profession has a long and, in degree at least, a respectable history. Certainly the Greeks and the Romans paid them a due regard. Julius Caesar, fulfilling his vow to the pirates who had captured him off the coast of Caria, cut their throats before he crucified them, thereby easing their passing – after a fashion. It was the sort of courtesy that would have been appreciated in New Providence in its heyday.

The Bahamas, despite the travel folders, are unlovely islands – low, often enough barren, and frequently carrying a spine of the ugliest broken coral limestone that may be found in a long voyage. Most of them are treeless, with here and there a cove furbished with feathery casuarinas, which come from Australia and look inappropriate off the coast of Florida. The mangroves of the swamps are like mangroves anywhere – sinister and hideous.

New Providence ('New' to distinguish it from Old Providence which Morgan was always capturing) is, for the Bahamas, a wide island, housing a small and undistinguished range of hills and two swamps – with mangroves.

The group is, of course, highly important because of San Salvador – and Christopher Columbus.

The Spaniards never settled there, they discovered the high – and more salubrious – islands sufficiently early. Charles II made a grant of them to his Attorney-General but nothing came of it, and very little at all happened to them before 1646, when William Sayle, Governor of Bermuda, sailed to them with the Eleutheran Adventurers.

The name is not entirely as romantic as it sounds. The original name of Eleuthera was, in fact, Isla de Terra, and the compromise is singularly English. But Eleuthera failed them, and the Adventurers moved to New Providence. The Lords Proprietors of Carolina acquired the Bahamas and handled them slackly. By 1676 New Providence was described as 'lewd and licentious', a condition to which it has reverted joyously at frequent intervals. Eighteen years afterwards it was so notorious as a pirate base that the Spaniards attacked it in a remarkable display of moral indignation – plus, of course, the hope of loot – and carried off the Governor.

In 1714 there were about 200 families scattered through Cat Island, Exuma, Eleuthera, Harbour Island and Abaco, with a handful at New Providence, 'without any face or form of Government, every man doing onely what's right in his own eyes'.

The situation was ripe for development.

In the same year the Plate Fleet with the season's treasure was caught in a hurricane and driven ashore on the Florida reefs.

The Spaniards acted with great energy. The wrecks were located, salvage teams were assembled, primitive diving gear provided, and a party of sixty soldiers landed to protect the salvage. Inevitably news spread rapidly. Everything was admirable except the security.

In 1716 at Port Royal, Henry Jennings, who had operated with moderate success as a privateer during the war of the Spanish succession, acquired precise information about the progress of the operation and opened a series of conversations with local merchants.

Jennings was not a buccaneer, buccaneering had been proscribed for half a century, but he had the instincts. The procedure at Port Royal was a carbon copy of the funding of a buccaneering venture. With rather more than 300 men in two ships and three sloops, Jennings sailed for the Gulf. The Spaniards had already dispatched to Spain some millions of pieces of eight, but Jennings's information was that 350,000 were currently stored in a hut on shore.

He achieved a masterly surprise, the landing was expertly executed, the guard bolted, and Jennings headed back for Jamaica with rather more than £90,000-worth of Spanish coin. He was, understandably, elated. So was his crew. When a Spanish ship was sighted, the transition from land robbery to sea piracy was made with no strain at all. They collected another 60,000 pieces of eight and a valuable cargo of cochineal, indigo, 'and many rich goods'. They were kind enough to leave the ship to the Spaniard but the cargo they transferred.

It sold well in Port Royal. Unfortunately, the Spaniard had followed them and reported their destination; there was a thunderous diplomatic reaction, and Jennings and the greater part of his crew decided, forthwith, to go 'on the account'.

A base was, of course, desirable. New Providence, 250 miles to the north-west, had all the advantages, as well as an all but perfect climate.

Who used its one village, Nassau, first is unimportant, it had been piratical for forty years; almost overnight now it became the pirate capital of the western Atlantic. Blackbeard (Teach or Tache), Bartholomew Roberts and Vane and Hornigold, Barrow and 'Calico Jack' Rackham, Walker and Sample and Israel Hands, Stede Bonnet and Harriot and scores whose names are lost to history sailed from it unscathed. Criticism of the Royal Navy became acrid. Its captains engaged in private trade, convoys were a matter of financial arrangement. 'Naval commanders', said the Governor of Jamaica, 'have no other regard than is consistent with their own private gains.'

It was the high noon of piracy.

It was also the pinnacle of the riotous splendours of Nassau. There were reputed to be 400 pirates ashore at any one time and at least another 500 at sea. Bay Street was a shanty town of brothels and gambling booths and drinking shops. There was little order and less law, and no drainage at all – it must have been a stinking Alsatia of the sea. But behind Hog Island and Athol there was anchorage for 500 small ships if needful, and there were two exits.

The glory was preposterously brief.

The 'Proclamation for Suppressing of Pirates' dates it from '. . . the 24th day of June, in the year of Our Lord 1715'. The proclamation itself was posted *in* Nassau in April 1718. The majority of the pirates that it was aimed against were in reality petty thieves, small time operators, casual criminals. Even among the great names there were few genuine leaders and no heroes. The history of the suppression proves it.

In 1717 the appointment by the pirates of two of their number, Barrow and Hornigold, as Joint Governors, with their offer of protection to all pirates, appears to have alarmed the British Government

– not altogether before time. Captain Woodes Rogers, the privateer who had captured the Acapulca Galleon, put forward a scheme for eliminating the nuisance. He was to be Governor himself, frigates were to be sent to cruise the area, and he was empowered to land with a force of two King's ships. A proclamation of amnesty was to be read, with full pardon for all pirates who should 'surrender to one of our principal Secretaries of State in Great Britain or Ireland, or to any Governor or Deputy Governor of any of our plantations beyond the seas' by 5 September 1718.

The proclamation relied, fundamentally, on the pusillanimity of the majority. But Woodes Rogers was a cautious man; he sent the proclamation in first – rather like putting one's hat around the door.

They took it, says Charles Johnson,

> . . . as Teague took the Covenant, that is they made prize of the ship and Proclamation too. . . . Some were for fortifying the Island, to stand upon their own terms and treating with the Government upon the footing of a Commonwealth.

The Proclamation, however, was insidiously attractive. It listed no penalties; the few men of comprehension among the pirates saw this early. Jennings, who had shrewdly evaded the Governorship but was 'Commodore' at the time, 'resolved to surrender without more ado'. When Woodes Rogers, properly escorted, followed the Proclamation in, only one ship, that commanded by Charles Vane, bolted. Vane 'slipped his cable, set fire to a large prize they had there, and resolutely put out, firing at the man-of-war as he went off'.

The rest followed Jennings's example. Rogers landed between files of penitent pirates who fired salutes over his head as he made his way to the Fort – which may have been interestingly nerve-racking.

The last phase of the Pirate Commonwealth is a sorry affair. Woodes Rogers acted fairly enough. He had endeavoured to set up trade to replace piracy and he tried to secure some sort of order in Nassau, but his reformed pirates relapsed alarmingly. Men he sent on distant voyages took up the account again, pirates at sea who had not come in threatened the colony. The long process of capture and trial and condemnation perforce began.

Woodes Rogers strung up ten of John Auger's crew in December 1718 – with suspect legality – but he found it difficult to obtain a proper backing. Finally, in 1721 he was recalled, and in London was imprisoned for debt.

Others took up the suppression.

On the whole the pirates died tamely: Burgess of disease on a trading voyage; Hornigold 'cast away upon rocks, far away from land'; Martel forced on an uninhabited island; James Fife killed by his own men. Israel Hands was actually sentenced to death, but a decision to prolong the pardon reached Virginia in time to save him and he got back to London 'to beg his bread'.

Teach – Blackbeard – was one of the few who died with a grim, ferocious gallantry. Caught at last in the James River by a party under Lieutenant Maynard of the sloop *Ranger*, he boarded the sloop and was wounded twenty-five times before he dropped dead in the act of firing a pistol. His head was severed and hung at the *Ranger*'s bowsprit end.

Bartholomew Roberts matched Teach in courage, though his action with H.M.S. *Swallow* was ill judged. He was killed by a grape-shot through the throat.

Vane and Sample and Cocklyn were hanged. In May 1722, forty-three men were 'turned off' from a single ship. The list is extensive, the catalogue of last dying words enormous.

Dennis Macarty, on the wall of the fort at Providence, said bluntly:

> I never thought to see the time when ten such men should be tied up and hanged like dogs, and four hundred of their sworn friends and companions quietly standing by to behold the spectacle.

Another of the same crew said:

> I do heartily repent. I repent I have not done more mischief, and that we did not cut the throats of them that took us, and I am extremely sorry that you aren't hanged as well as we.

Even 'Calico Jack' was eventually tried and hanged, and the night before his execution Anne Bonny, his wife, went to him and whispered scathingly

> . . . that she was sorry to see him there, but if he had fought like a man, he need not have died like a dog.

Anne Bonny and Mary Read, her friend and fellow pirate, still deserve adequate analysis and a proper space in the literature of psychiatrics. Transvestism, lesbianism, sadism, enthusiastic sexual normality, and side issues of homosexuality weave an intricate pattern through their stories.

Mary Read's mother was the London wife of a merchant seaman. She bore him a child and, in his absence, suffered the occupational hazard of merchant seamen's wives and conceived again. To keep her secret from his family, she withdrew from London. The son died; the second child, a girl, Mary, lived. Some considerable while later, having run out of money, she returned to her mother-in-law, dressing Mary as a boy, managed, incredibly, to pass her off as the son, and asked for and got an allowance to support him – her husband having disappeared.

In due course the grandmother died, and Mary was hired to a French woman as a footboy. Grown up, she deserted the back stairs and served as a seaman on a man-of-war. She left the sea in due time and joined a regiment of foot in Flanders as a cadet. Disappointed of a commission, she joined a cavalry regiment, did well in action and then, disastrously, fell in love with a young Fleming in the regiment. Coyly, Johnson says in the *General History of the Most Notorious Pirates*: 'She found a way of letting him discover her sex, without appearing that it was done with design.' Having done so, she held out on him until he agreed to marriage.

The regiment seems to have taken it remarkably well. The couple were set up in an eating house near Breda called The Three Horseshoes, and prospered. But her husband died, peace came, and she returned to her disguises. Eventually a ship in which she was bound for the West Indies was taken by pirates, and she slipped with complete ease into a new trade.

She was one of those who accepted the Proclamation, and when Woodes Rogers commissioned privateers, she went to sea again with 'Calico Jack' Rackham. When Rackham decided to go once more 'on the account', she continued with him.

Rackham's mistress, Anne Bonny, was born in Ireland. With her father she moved to the Carolinas where she had a turbulent childhood. In man's clothes, she bolted from her husband with Rackham, and took up the pirate lay.

In due course she fell in love with Mary Read, to Rackham's entirely understandable fury. In the meantime – the story has the complexities of a Victorian novelette – Mary was being harassed by an amiable young man who had conceived a homosexual passion for her.

To attempt to unravel the affair is fortunately unnecessary; it was settled externally. Rackham, raiding without due care along the north coast of Jamaica, delayed too long, was overtaken by a sloop sent after him by the Governor 'and after a very small dispute' – the gravamen of Anne Bonny's caustic farewell – was captured, taken to Port Royal and tried.

Rackham, Mary Read, Anne, and eight members of the crew were sentenced to death. Three of them – Rackham, and two men named Featherston and Corner – were hanged in chains. Anne Bonny and Mary Read, in the ancient formula, 'pleaded their bellies'. Both were fortunately pregnant.

It was not the end of piracy at Providence or, for that matter, of piracy in the West Indies, still less of piracy and islands. It is, as has been said, a very ancient and it has been, at times, a respectable profession. It is capable of much endurance as long as islands remain.

Index